WELLBEING AND PLACE

Wellbeing and Place is a wonderfully wide-ranging examination of two increasingly important policy concepts. This must-read collection offers a critical analysis of mainstream approaches to wellbeing, and shows convincingly how wellbeing and space are linked insolubly. This book takes our understandings of wellbeing in new and exciting directions.

Tim Schwanen, University of Oxford, UK

Wellbeing and Place

Edited by

SARAH ATKINSON, SARA FULLER and JOE PAINTER

Durham University, UK

Routledge

Taylor & Francis Group

LONDON AND NEW YORK

First published 2012 by Ashgate Publishing

Published 2016 by Routledge
2 Park Square, Milton Park, Abingdon, Oxfordshire OX14 4RN
711 Third Avenue, New York, NY 10017, USA

First issued in paperback 2016

Routledge is an imprint of the Taylor & Francis Group, an informa business

British Library Cataloguing in Publication Data
Wellbeing and place.
  1. Well-being. 2. Human ecology. 3. Human beings--Effect
of environment on. 4. Environmental psychology.
  5. Environmental health.
  I. Atkinson, Sarah. II. Fuller, Sara. III. Painter, Joe,
  1965-
  304.2'3-dc23

Library of Congress Cataloging-in-Publication Data
Wellbeing and place / [edited] by Sarah Atkinson, Sara Fuller, and Joe
Painter.
      p. cm.
  Includes bibliographical references and index.
  ISBN 978-1-4094-2060-6 (hardback)
  1. Place attachment--Psychological aspects. 2. Human
beings--Effect of environment on. 3. Well-being I. Atkinson, Sarah. II.
Fuller, Sara. III. Painter, Joe, 1965-
  BF353.W45 2012
  155.9'04--dc23

2012002351

ISBN 13: 978-1-138-25474-9 (pbk)
ISBN 13: 978-1-4094-2060-6 (hbk)

Contents

List of Figures

List of Tables

List of Contributors

Editors

Sarah Atkinson is an Associate Director of the Centre for Medical Humanities at Durham University and a Reader in the Department of Geography. She studied Anthropology and Nutrition and has worked in development and public health, both in research, management and consultancy roles. Her current research examines the role of health and medicine in human flourishing through interrogating the nature of caring practices, the politics of wellbeing and the spatialities of arts, creativity and imagination.

Sara Fuller is a Research Associate in the Department of Geography at Durham University. She joined the department following completion of her doctoral research at the University of Sheffield which explored environmental protests in relation to large scale infrastructure projects in Central and Eastern Europe. Her current research focuses on concepts and practices of justice and democracy in the field of the environment. Prior to gaining her PhD, Sara worked as a researcher at the Centre for Sustainable Development, University of Westminster exploring the interfaces between social exclusion and the environment.

Joe Painter is Professor and Head of the Department of Geography at Durham University, Durham, UK. He holds a BA in Geography from Cambridge University and PhD in Geography from The Open University. His academic expertise lies in the field of political geography and he is the co-author with Alex Jeffrey of *Political Geography: An Introduction to Space and Power* (Sage, 2009). He has published widely on urban and regional politics and government, theories of territory, and the geographies of citizenship and the state. He has a particular interest in the relationship between the state and everyday life.

Contributors

Helen Beck currently works for the UK Drug Policy Commission, conducting research into the impact of increasing public sector austerity, alongside greater localism, on action to tackle the problems associated with illicit drugs. Prior to this, Helen worked as a Public Space Research Advisor at the Commission for Architecture and the Built Environment (CABE). For five years, she managed research looking at the value green and open spaces contribute to urban areas.

This programme of work aimed to make best use of existing data to measure the 'state' of urban green space in England and examined the impact of green space quality on the health and wellbeing of residents in deprived areas, with a specific focus on ethnicity. Helen has also worked as a Research Officer for the London School of Economics. She holds a MA in Development and MSc in Housing and Regeneration. Her research interests include understanding the role of good quality spaces in promoting positive health and wellbeing in low income areas, the impact of government policy on areas and the structures and processes that create successful and sustainable communities.

Melania Calestani trained as both a geographer and anthropologist in Italy and the UK, Melania has carried out research in the Bolivian highlands, Western Samoa, Guam, Europe and the USA. Her main interests include issues related to wellbeing, faith, morality and value. Her work has mainly focused on individual and collective definitions of 'the good life', exploring potential contradictions between different models and analysing how individual dimensions compete with collective ones in this context. She has used qualitative and participatory methods to explore the possibility of developing better wellbeing indicators. Her research has also engaged with themes of political economy and material culture, especially in relation to commodity fetishism, the social life of things, waste and excess. Melania is currently a Research Fellow at the University of Southampton where she is involved in a research project about health inequalities and quality of life in the UK.

Sandra Carlisle has an academic background in medical social anthropology and sociology applied to public health and community health development. Over the years, she has conducted numerous public health and policy-related research and evaluation projects, including ethnographic studies of GP care for older people, informal community support for family carers, and community-based action research and partnership work on health inequalities. She has worked at Dundee University, University of Wales College of Medicine, the University of Edinburgh, and Keele University. Currently a research fellow at the Institute for Health and Wellbeing at the University of Glasgow, she is exploring the relationship between 'modern culture', wellbeing, and larger global problems. Much of this work is publicly available through a dedicated website (www.afternow.co.uk).

David Conradson is Senior Lecturer in Human Geography at the University of Canterbury, New Zealand. His research focuses on wellbeing and resilience in western societies, drawing on the perspectives of social, cultural and health geography. His thinking is also shaped by work in psychology and psychotherapy, community and public health, and by interdisciplinary scholarship on emotion and affect. Before joining the University of Canterbury, he worked at the University of Southampton and was a Visiting Fellow at Durham University in the UK. He currently serves on the editorial advisory boards of *Social & Cultural Geography* and *Social Science & Medicine*.

Sarah Curtis is Professor of Health and Risk at the Department of Geography and Institute of Hazard, Risk and Resilience at Durham University, UK. She is an internationally-renowned scholar in the geography of health focusing on the geographical dimensions of inequalities of health and health care. Her scholarship elucidates how, and why varying geographical settings relate to human health inequalities. Her work has strong applied as well as theoretical aspects, contributing to health policy development and evaluation of health services in the UK, France, Russia, Poland, Canada and the USA. Recent research includes: adaptation of the social and care services of older adults in the context of climate change; health impact assessment of urban regeneration schemes, (for the Department of Health, and other agencies) and development of healthy public policy (with agencies in Canada and UK).

Lorraine Gibson is a Research Fellow at the Centre for Research on Social Inclusion at Macquarie University and a Research Associate with the Australian Museum in Sydney. Her work with Aboriginal and settler Australians engages with questions of race relations, cultural identity and social placement. As a social anthropologist, her methodology is grounded in the ethnographic however, her theoretical interests and thinking is interdisciplinary. Lorraine is currently exploring the place of 'class' in Australian Indigenous societies and how different cultural understandings and approaches to (un)employment, 'work' and social placement intersect with race relations, personal aspirations and cultural identity. Lorraine currently serves on the editorial advisory board of the *Australian Aboriginal Studies* Journal.

Rose Gilroy is Senior Lecturer in the School of Architecture, Planning and Landscape, Newcastle University. She joined academia from a housing practice background. Rose's research explores the impact of the ageing population, particularly in relation to planning and housing. Her concerns embrace the home and the lines of attachment that radiate from it into neighbourhood, city, region, maybe even cyberspace but also burrowing down into home to think about favourite spaces within it and cherished objects. In the wider arena, Rose explores how place supports everyday life in later life or not of course. As a planner, she is interested in the transactional relationship between people and their environment that embraces how people individually and collectively may influence their environments. Within the container of home she reflects on how this place that we experience physically, socially, emotionally contributes to the persistent existential question – who are we? Rose's research has ranged from work in the north east of England to urban China and rural Japan.

Phil Hanlon was educated in the West of Scotland and graduated in medicine from Glasgow University in 1978. Following a period of clinical experience in adult medicine and general practice, he took up a research post with the Medical Research Council in the Gambia. On returning to the UK, he completed a period of training in public health after which he was appointed to the post of Director of Health Promotion for the Greater Glasgow Health Board. In 1994, he became a Senior Lecturer in Public health at the University of Glasgow and was promoted to Professor in 1999. Between January 2001 and April 2003, Phil undertook a secondment to establish the Public health Institute of Scotland, before returning to full-time academic life as Professor of Public Health at the University of Glasgow. Since then he has been involved in numerous research projects, and leads the 'Culture and Wellbeing' study, funded by the Scottish Government. Other research interests include the uses of integrated public health data and evaluation of complex public health interventions. Most recently, he has been actively engaged in developing new forms of thinking for public health.

Gregor Henderson works as a consultant and advisor to a number of Government Departments, both in the UK and overseas, working with public sector, not for profit, community based and private sector agencies on mental health and wellbeing. Gregor is currently working as an adviser for the Department of Health on wellbeing and public mental health. From 2003-2008, Gregor was the first Director of Scotland's innovative and internationally acclaimed National Programme for Improving Mental Health and Wellbeing. Prior to that he was the founding Director of the Scottish Development Centre for Mental Health and a Senior Fellow at the Institute of Psychiatry, King's College, London. Gregor is also a policy adviser to Mental Health Europe and an advisor to ThePlace2Be, a national UK children's school based mental health charity. Gregor is interested in combining policy, research, evidence, practice and people's lived experiences to transforming the way people, families, communities and societies think and act about mental health and wellbeing.

Gordon Jack is Professor of Social Work (Children, Youth and Families) at Northumbria University. Following professional qualification as a social worker and 15 years as a practitioner and manager in local authority children's services, he took up his first academic post in social work in 1991. His main research interests lie in the fields of social ecology, social policy and child wellbeing, in particular applying an ecological framework to investigate the effects of poverty, inequality and disadvantage on communities, family functioning and child wellbeing. Most recently, this has involved investigating the influence of children's attachments to the places in which they have grown up on the development of their identity and their sense of security and belonging, with a particular focus on what happens to children brought up in care, who have experienced significant disruptions to their place attachments. These issues have major implications for policy and practice in child welfare, particularly in relation to assessments of need and risk, the design of interventions that aim to enhance the social capital and community cohesion of disadvantaged neighbourhoods, and the social support networks and place attachments of parents and children. Some of this work has had a significant influence on government guidance for social workers and other welfare professionals in children's services.

Jo Little is Professor of Gender and Geography at the University of Exeter. Her main research and teaching interests are in rural communities and social identities and in gender and geography. She has recently been involved in research on women and the rural labour market looking at the constraints that impact on women's access to employment in the countryside and the relationship between the cultural construction of rurality and women's involvement in paid work. She has on-going research on gender, safety and fear of crime in rural communities. Her research interests in gender relations are currently being developed in studies on the construction of rural masculinities and femininities and on the rural body. Jo has also worked on local food and rural regeneration in the South West and West Midlands of England. The majority of Jo's research has been in the UK although she is also involved in work on rural New Zealand.

Andrew Lyon studied sociology and economics at Edinburgh University (1976-1980). He conducted research for his PhD in India between 1981-1983 and again in 1985. In 1986, he became co-ordinator of the Glasgow Healthy Cities Project, part of the WHO Europe programme. During this period, he also acted as a consultant for the WHO in Europe and Bangladesh. He moved to Forward Scotland, the sustainable development charity, becoming acting chief executive in 2000. He left in May 2001, joining the International Futures Forum (IFF) as Converger. Andrew has engaged in voluntary activity for most of his adult life, including a spell as the Scottish Director of Sustrans, the cycle path charity. He was, until recently, a founding director at Common Wheel, a charity, which helped rehabilitating mental health patients to recycle bicycles at prices people can afford. Currently he is a director of Community Renewal, a charity that supports people living in deprived circumstances to fulfil their aspirations, and of AdHom, a medical charity based in Glasgow. He has also worked as honorary lecturer in Public Health at the University of Glasgow.

Stuart Muirhead is currently a Research Associate at the Public Services Academy within the University of Sheffield. He completed his doctoral studies in human geography at the University of Dundee where his thesis explored the links between well-being, nature and environmental volunteering, with a focus on exploring how social and emotional capital can be engendered through active nature work. There is an expanding body of research on environmental volunteering and the impact it can have on health and well-being and this project engaged with, and developed, these debates. Stuart was an active member of the post-graduate research community of geographers in the UK and co-convened two sessions on voluntary and community action at the international IBG/RGS conference in Manchester in 2009. Stuart has carried out research projects with Volunteer Development Scotland and the Scottish Agricultural College and, following completion of his PhD in 2011, also spent six months working within the Social Science and International Research Unit of the Scottish Government.

David Reilly (FRCP, MRCGP,. FFHom) is currently the Scottish Government's National Clinical Lead for Integrative Care; Consultant Physician, The Centre for Integrative Care, Glasgow Homoeopathic Hospital; Honorary Senior Lecturer, Glasgow University; Visiting Professor of Medicine, University of Maryland; faculty member at Harvard Medical School, USA; Director AdHominem Charity. He is a clinician, teacher and researcher who became a doctor to combine head and heart in helping people. En route to his training as a physician and GP he evaluated aspects of complementary medicine (including RCTs of homoeopathy) and mind-body medicine. He has wondered about better ways of approaching medicine and human caring that emphasize the innate healing capacity in people, the factors that modify the healing response, and their interaction in the therapeutic encounter and relationship. Currently he is exploring how knowledge from the study of healing process and wellness enhancement might be scaled up: e.g. in one-to-one relationships; in group work – like The WEL programme; in integrative models of care; in healing environments; and cultural and national development.

Mylène Riva is a Banting-Canadian Institute of Health Research Postdoctoral Fellow at Axe santé des populations et environnementale at Centre de Recherche du Centre hospitalier universitaire de Québec, Canada. Mylène holds a Bachelor's degree in Geography (2002) and a PhD in Public Health and Health Promotion (2008) from the University of Montreal, Canada. Between 2008 and 2011, Mylène was a Postdoctoral Research Associate at the Department of Geography and Institute of Hazard, Risk and Resilience at Durham University, UK. Mylène's work examines geographic and social inequalities in health and health-related behaviours and the wider determinants of these inequalities, with a focus on health in rural and remote settings and resource-reliant communities. Recently published work includes investigations of the links between the material and social conditions of geographic life environments and how these relate to physical and psychosocial health outcomes in Canada and the UK.

Rebecca Schaaf is a human geographer with research and teaching interests in development and sustainability. Rebecca's doctoral research used a wellbeing framework to explore the processes and outcomes of community group membership in a village in Northeast Thailand. The research was prompted by government promotion of community groups and microfinance facilities as a way of achieving the development goals of social cohesion and increases in individual wealth and wellbeing. Rebecca is currently a Senior Lecturer at Bath Spa University and her research and teaching explore various aspects of the development process, from issues of sustainability and environmental change, to the politics of collective action, poverty alleviation and enhancing wellbeing. She is also involved in research focusing on the teaching of development geography and methods of supporting students while on overseas and work-based learning placements. Other research interests centre around development in Southeast Asia, sustainable development, and the role of complexity theory in the social sciences.

Karen Scott has a PhD in Politics and in 2010 began a five-year research fellowship at the Centre for Rural Economy at Newcastle University. Her doctoral study focussed on the relationship of wellbeing, sustainability and democracy at the local governance level. This work led to her book *Measuring Wellbeing: Towards Sustainability?* (Earthscan by Routledge 2012) in which she explored the role of knowledge and power in developing national and local measures of wellbeing. Prior to becoming an academic she worked for several years in the voluntary sector in Edinburgh and Newcastle with charities supporting homeless people and offenders. She has also worked with communities regarding environmental conflicts over water resources, pollution and flooding and is trained in participatory appraisal. Her experience in social care, community planning and environmental fields underpins her research interest in the politics of human wellbeing and sustainable development at the local level. She is also currently exploring new research directions related to the role of creative economies in promoting wellbeing and sustainable development in rural areas.

Andrea Wheeler is a researcher at Loughborough University engaged in multidisciplinary research. Her research history and experience spans mechanical engineering, through design, to architectural philosophy and gender politics. Her current social sciences research in post occupancy assessment of new school buildings and the problems of encouraging sustainable lifestyles demonstrates her ability to work across interdisciplinary divides and to address complex science and humanities issues. She held a prestigious Research Fellowship from the Economic and Social Sciences Research Council to explore the role of schools design in building sustainable communities (2007-2010) and has worked for Defra's Sustainable Behaviours Unit (2010). Currently her research assesses the difference between predicted and actual energy performance of new school buildings. Andrea has presented her research at a number of international conferences.

Ian Wight is an Associate Professor, City Planning, Faculty of Architecture, University of Manitoba, Winnipeg, Canada. He has three degrees in Geography (MA Hons and PhD from Aberdeen; MSc from Alberta) and was a professional planner in Alberta and BC before assuming his academic position. Ian is a founding member of the Integral Institute, associated with the Integral Ecology branch since its inception. Ian has been a member of the Canadian Institute of Planners since 1977, and has taught the capstone Professional Practice course at the UofM for 16 years. His current interests include the inter-relationship of place, place-making and planning, and 'evolving professionalism beyond the status quo', both from an integral perspective. The material presented here was largely developed while on leave, first at the Centre for Human Ecology, and then at the Centre for Confidence and Wellbeing, both based in Glasgow, Scotland.

Foreword

Wellbeing and place are inextricably inter-connected. Our economic and educational opportunities, our social and political relationships, our environment, and our imaginative, cognitive and creative worlds are all profoundly impacted by where we live and who we live alongside. However, our access to the resources required to generate wellbeing, whether environmental, collective or personal, exhibit huge spatial inequalities in their distribution; between different regions of the world, between countries, within countries and between local areas and even within households. This is increasingly recognized with wellbeing becoming an explicit public policy goal in many countries, including Canada, France and the United Kingdom, and measures of wellbeing are being incorporated into some national systems for assessing progress. Wellbeing can be conceptualized and defined in a variety of ways and, as such, its use within policy is highly political. Given the concurrent focus on place-making and place-shaping within local governance arrangements, there is a pressing need for critical attention to such policy formulation and an examination of the multiple ways in which wellbeing and place mutually constitute one another.

This volume, which emerges from an international conference organized by Durham University in 2009, presents a significant and pioneering collection in addressing this endeavour. It brings together provocative contributions from leading researchers working in different academic disciplines and combines both applied and more critical perspectives. It advances the academic study of wellbeing and place, whilst also offering insights into how policymakers can improve the wellbeing of people and places.

Professor Clare Bambra,
Durham University, UK

Chapter 1

Wellbeing and Place

Sarah Atkinson, Sara Fuller and Joe Painter

Serious moves are afoot to shift the assumptions of governments and populations alike away from the idea that a flourishing life is primarily connected to material prosperity and towards positioning wellbeing as the ultimate goal for policy intervention (NEF 2004, Stiglitz et al. 2009). At the same time, there has been an ideological and rhetorical urge towards more responsive policy-making through greater local voice, local accountability and decentralized governance, a shift mobilized through an emphasis on the processes of place-making and place-shaping (Shneekloth and Shibley 1994, Steuer and Marks 2008, Wight this volume). These current shifts to reconfigure policy landscapes are predominantly in high income countries but intersect with existing debate over the aims of development interventions (Gough and McGregor 2007) and the impacts of both global and local inequalities (Singh-Manoux et al. 2005, Wilkinson and Marmott 2003, Wilkinson and Pickett 2009).

The centrality of two concepts, wellbeing and place, within contemporary governance and policy makes it timely to examine their possible meanings and the relationships between them through a range of academic inquiries that can offer policy relevance and critical reflection. This volume offers a collection of chapters from a range of disciplinary and policy-engaged directions arranged under two broad categories: those determining wellbeing's relationship to place and those contesting the definitions and relationships of wellbeing and place. In framing an introduction to the collection, we argue that there is a dominant approach in wellbeing research and policy. This approach is reflected in research on wellbeing and place which can contribute in a relatively straightforward manner to current policy agendas and debate. This does not imply that such research cannot deliver trenchant critique or expose injustices in policy formulations and implementation (see for example, Jack this volume); on the contrary, research that can connect to dominant framings is more likely to have an impact on policy and see such critiques of formulation and implementation reflected in practice (see for example Beck this volume). However, there is a second body of research which interrogates the assumptions within a dominant approach. There is a wide range of theoretical and methodological perspectives from which this more conceptual critique may derive. Empirically informed research on differently situated understandings and experiences of wellbeing expose situated conflict and the power inequalities in terms of which perspectives are valued (see for example, Gibson this volume). Social theories provide critical analytical insights through which to interrogate further these highly situated relations of power (see for example, Little this volume).

The central argument in the emergence of wellbeing as a governing policy concept is that economic growth and material wealth should be seen as the means to a good life rather than the end itself. This argument has emerged through movements related to various political concerns including those addressing broad development goals (Anand and Sen 1994, Nussbaum 2000), environmentally sustainable living (Defra 2005, NEF 2004) and a focus on an individualized, psychological state of happiness or flourishing (Layard 2005, Seligman 2011). Whilst these different routes to a common call for policy to look beyond economic measures of social progress are all captured under a label of wellbeing, the various political engagements with the notion of wellbeing involve very diverse conceptualizations in terms of scale, scope, location and responsibility (Atkinson and Joyce 2011, Ereaut and Whiting 2008, Sointu 2005). A similar diversity of conceptualizations and methodology is reflected in the engagement with wellbeing across different academic disciplines (de Chavez et al. 2005).

An intellectual inquiry into what it is that makes for a good or flourishing life is nothing new. Scholars of wellbeing in Western Europe typically trace the contemplation of individual wellbeing back to classical Greece and to the competing philosophies of hedonic, or happiness and pleasure-based, wellbeing of Aristippus (Kahnemann et al. 1999) and eudaimonic, or satisfaction and meaning-based, wellbeing of Aristotle (Deci and Ryan 2008). The hedonic pathway can be tracked through later philosophical contributions of Mill and Bentham on how to select between alternative individual and collective actions in order to maximize the greatest happiness or utility for all. In recent times, psychologists working within the hedonic tradition have provided high profile and policy-relevant research including the contributions to a social policy of Ed Diener (2009) and the more personalized approaches to happiness and wellbeing of Martin Seligman (2002, 2011, see Table 1.1). The engagement of the economist, Richard Layard, with the concept of happiness and its determinants has been highly influential through providing renewed economic credibility to the concept (Layard 2005). Research showing that positive and negative affect constitute independent dimensions adds further intellectual complexity to understanding how an overall balance between these may be reached (Huppert and Whittington 2003). The eudaimonic pathway can similarly be tracked to its contemporary expression in both psychological and more economic fields. In psychology, this approach attends to the dynamic processes that enable and re-enable a sense of self-fulfillment, meaning and purpose (Deci and Ryan 2008); an influential approach of this kind is the six characteristics of psychological wellbeing defined by Ryff (1989; Ryff et al. 2004, see Table 1.2). The critical engagement with the policy intentions of development interventions, issues of social justice and inequity associated with the capabilities approach (Nussbaum 2000, Stiglitz et al. 2009, see Table 1.3) and the Wellbeing in Development (WeD) group (Gough and McGregor 2007) and the Wellbeing and Poverty Pathways collaboration (2011) offer broader engagements with wellbeing which combine objective and subjective assessment and cover material, relational, cognitive, affective and creative dimensions to wellbeing. However, whilst these different approaches offer a richness of perspectives and encounters with what might constitute a flourishing life, they are all challenged by the contemporary concern with environmental sustainability to incorporate a greater temporal dimension which can consider not just the sustainability of wellbeing within an individual life-span, but equally, if not more importantly, beyond and into future lives, of both human and other co-present species (Defra 2005, NEF 2005).

Wellbeing, however defined, can have no form, expression or enhancement without consideration of place. The processes of well- being or becoming, whether of enjoying a balance of positive over negative affects, of fulfilling potential and expressing autonomy or of mobilizing a range of material, social and psychological resources, are essentially and necessarily emergent in place. And, as with wellbeing, engagements with a concept of place have provided a rich field of debate and contestation. Whilst some quantitative and policy-relevant approaches treat place as little more than a static backdrop or a container against or within which social interactions occur, others, particularly from the field of human geography, have problematized this apparently common-sense approach to position place as inherently relational in both its production and its influence (Cresswell 2004). Contemporary policy for local governance reflects these more complex mobilizations of a relational place, as demonstrated through the importance increasingly given to processes that help to build a sense of community, social cohesion and trust, enhanced liveability and so forth, an approach which in some countries has come to be labelled as place-making or place-shaping (Mullins and Van Bortel 2010, Gallent and Shaw 2007, Wight, this volume). In these policy formulations, whilst place-making is an important outcome of local government, in relation to wellbeing it is clearly positioned only as a determinant of personal wellbeing, aggregated for the constituent population, rather than as an expression of a more collective or temporal wellbeing in its own right.

Policy-facing research on wellbeing tends to be embedded within an accepted line of argument that calls for greater clarity in definition and use of the term wellbeing. The argument is that first, we lack agreed terminology, definitions and monitoring tools, secondly that this is important because different understandings and definitions of wellbeing risk creating barriers to communication across different sectors involved in policy-making and thirdly, that in order to evaluate the benefits of different policy interventions in terms of enhancing wellbeing, standardized indicators and monitoring tools are essential (Ereaut and Whiting 2008). The co-existence of several discourses of wellbeing is evident within current policy debates (Atkinson and Joyce 2011, Ereaut and Whiting 2008) and the specifics of how wellbeing is defined in terms of the dimensions used and how these may be weighted in different approaches show great variation. However, despite such variation, there is nonetheless a considerable degree of convergence in the underlying approach taken in defining both wellbeing and place across policy documents and policy-facing research which can be seen as an emerging dominant approach in operationalizing wellbeing and place.

First, research mostly takes an approach to definition that deals with the abstract nature of the concept by breaking it down into constitutive dimensions in what we are calling a components approach to wellbeing. A components approach underpins economic and psychologically informed schemes of wellbeing and both hedonic and eudaimonic variants. Tables 1.1-1.3 provide examples of some of the well-known frameworks for wellbeing that illustrate the components approach. Component lists have also been derived from empirical research on people's own definitions of what is important to their wellbeing and one of many possible examples is the framework produced by the Office for National Statistics in the United Kingdom to shape a national consultation on wellbeing which was structured through a set of domains derived from preliminary public engagement (ONS 2011).

Table 1.1   Examples of Psychological, Subjective and Hedonic Components

Layard 2005; Steuer and Marks 2008	Veenhoven 2000	Seligman 2011
Single-item life satisfaction 'All things considered, how satisfied are you with your life as a whole?'	Liveability of environment Life-ability of individual External utility of life Inner appreciation of life	Positive emotion Engagement, interest Relationships Meaning Accomplishments

Table 1.2   Examples of Psychological, Subjective and Eudaimonic Components

Ryff 1989	Ziegler and Schwanen 2011 (empirically derived)	ONS national consultation framework 2011 (empirically derived)
Self-acceptance Autonomy Personal growth Environmental mastery Purpose in life Positive relationships with others	In relation to ageing: Physical health Independence Mental health and emotional wellbeing Social relations Continuity of self and self-identity	Relationships Health What we do – work, leisure and balance Where we live Personal finance Education and skills Contextual domains – governance, economy, natural environment

Table 1.3   Examples of Economic and Developmental, Objective and Subjective and Eudaimonic Components

Stiglitz et al. 2009	Nussbaum 2000	Clarke 2006
Material living standards	Life	Basic – calorie intake/day,
Health	Bodily health	access to safe water
Education	Bodily integrity	Safety – IMR, life
Personal activities	Senses/imagination/thought	expectancy
Political voice and	Emotions	Belonging – telephone
governance	Practical reason	mainline, fertility rates
Social connectedness	Affiliation	Self-esteem – adult literacy,
and relationships	Other species	unemployment
Environment	Place	
Security	Control over one's	
	environment	

A second feature of the dominant approach to wellbeing is that all these endeavours share a common understanding of wellbeing as a quality that inheres to the individual. The scope of wellbeing may range from an inner balance between positive and negative affect through to a breadth of components and it may be influenced by factors and processes from proximal personal interactions through to global scale processes. Wellbeing may be assessed objectively or subjectively, as a snap-shot of a current state, longitudinally across time or as a projection into the future, but in all these diverse scenarios, the central concept of wellbeing is itself individual in scale. Wellbeing was not always conceived of as an individual quality; earlier outings of the concept addressed collective aspects of the good life in terms of the economic wellbeing of the nation (Sointu 2005) or the moral landscapes that may inform or confront social and environmental injustice (Smith 2000). These aspects are now largely positioned as contextual influences on individualized wellbeing. Community or population measures of wellbeing exist in contemporary usage as aggregates of individual measures. Nonetheless, despite the individualization of wellbeing, there are still alternative, less dominant discourses of wellbeing in current policy communities that treat the concept as collective, most prominently in relation to sustainability and environment (Atkinson and Joyce 2011, see Wheeler this volume). Indeed it is the parallel existence of different mobilizations in different policy sectors that has given rise to calls to decide on an agreed terminology (Ereaut and Whiting 2008). The third feature of a dominant approach to wellbeing is that, despite variation in component sets as indicated in Tables 1.1-1.3, there has been a marked tendency for wellbeing to be used as a synonym for health, both in research and in policy (Atkinson and Joyce 2011; Ereaut and Whiting 2008; WHO 1948; see also Riva and Curtis this volume). Moreover, this conflation goes even further in that not only is wellbeing reduced to health, but particularly to mental health, to the absence of mental ill-health and increasingly to the concept of personal resilience (Atkinson 2011).

Debate in establishing sets of such individually attributed components centres on identifying which components are essential in defining wellbeing and which comprise the influences that produce wellbeing. Defining the essential set for wellbeing requires the specification of components that operate largely independently from one another and decision as to whether such components are best captured through objective or subjective assessments. The next step from these more conceptual elaborations of the nature and components of wellbeing is to identify associated, presumed influential, variables usually through quantitative research designs. In this there is a veritable burgeoning industry in exploring the determinants of wellbeing, from individual factors through to the global. Certain aspects of human life commonly feature in such explorations, including social relationships, health, safety and financial security, some variant of control in life, some variant of meaningful purpose, but, depending on how wellbeing is conceived initially, these may be cast as either dimensions to wellbeing or determinants of wellbeing. Thus in a hedonic psychological approach to a personal subjective wellbeing, economic and various other social elements are cast as determinants; this is the case in the Easterlin paradox, the much cited and influential proposition that happiness has not improved over time despite evident gains in material prosperity (Easterlin 1974; 1995, Layard 2005; see also Albor 2009 and Stevenson and Wolfers 2008 for a critique of this argument). By contrast, a eudaimonic, economic or developmental approach attempts to define the entirety of human flourishing through independent dimensions to wellbeing which cover a wide range of both objective and subjective components (Nussbaum 2000, Stiglitz et al. 2009).

There has been a vast quantity of research based on this kind of approach to wellbeing. In this, the fields of human geography and planning have contributed by their specific engagement with spatial relations. For example, Brereton et al. (2008) demonstrate how the inclusion of spatial variables into models exploring variation in subjective, hedonic wellbeing in relation to environmental characteristics can increase the explanatory power by a factor of three. Nonetheless, this engagement has, to date, predominantly focused on wellbeing as health, particularly within human geography (Collins and Kearns 2005, Cummins and Fagg 2011, Eyles and Williams 2008; Groenewegen et al. 2006, Kearns and Reid-Henry 2009, see also Conradson this volume for a review of the literature). New policy trends in governance towards place-making and the greater prominence of concerns with wellbeing are likely to generate indicators of wellbeing that can be disaggregated to units of local government and most likely to neighbourhood scales (NEF 2010). However, similar to the academic work within human geography, much of the existing planning and policy-facing research on wellbeing that has fore-grounded the specific influences of place and place-related variables has predominantly invoked a health-related interpretation.

The importance of subjective engagement with local environments has emerged strongly in research and planning in contrast to the existing convention of describing and assessing environment through objective measures. Michael Pacione (2003) has stressed this in respect to urban planning, while qualitative, empirical research with local residents has brought to light the importance of individual biographies in mediating and managing negative aspects of local environments (Airey 2003). The significance for wellbeing of subjective, emotional investments in objective, material markers of economic wellbeing is well illustrated in a study of investment in housing. Those who have made an explicit financial investment into their housing tend to score lower on a subjective wellbeing scale than more generalized investors. The latter group, being relatively relaxed about the return on their housing investment, take greater enjoyment from the affective and practical value of their housing (Searle et al. 2009). Whilst much of the research structured through the dominant approach to wellbeing treats place as something of a given, a static snapshot of settings in which various resources and experiences influence both objective and subjective wellbeing, the attention to subjective wellbeing inevitably draws in questions of what places mean to people and the role of emotional attachments to places (see Jack this volume). Such recognition is difficult to combine with a mobilization of place as contextual backdrop but rather insists on a reconceptualization of places as profoundly relational. Human geographers, for example, have described the mutual constitution of emotion and place as an emotio-spatial hermeneutic in which, 'emotions are understandable – 'sensible' – only in the context of particular places. Likewise, place must be felt to make sense' (Davidson and Milligan 2004: 524). The complexities and multiple interconnections in engendering personal and collective expressions of a situated and relational wellbeing, whether subjectively or objectively defined, are difficult to capture through the dominant approach. Alternatives include the so-called spaces of wellbeing approach from human geography (Fleuret and Atkinson 2007, Hall, 2010, Herrick 2010; Smith and King 2009) which proposes that wellbeing is emergent through four inter-related spaces of resource mobilization: capabilities, social integration, security and therapeutic processes. This resonates with subtle and detailed case studies and ethnographic research in which the complexities of wellbeing are comprehended as produced within multiple assemblages of the material, cultural, emotional and social constituting meaningful sites of living and encounter (Panelli and Tipa 2009).

This volume of chapters brings together a range of perspectives on wellbeing across different disciplines including those addressing immediate applied policy concerns as well as those offering more critical academic engagements. The chapters have been grouped into two sections: those that explore the dynamics that determine wellbeing in relation to place and those that explore contested understandings of wellbeing both empirically and theoretically. In keeping with a dominant approach, the chapters mostly engage with wellbeing as a personal, individual attribute and largely address health and psychological dimensions to wellbeing. The majority of chapters relate to high income and Anglophone settings, particularly, although not exclusively, to the United Kingdom. However, care and attention to the politics of different wellbeing perspectives and their intersections with place can provoke challenges to a dominant approach within these settings as much as from the small number of chapters that reveal alternative perspectives perhaps more overtly.

The first part comprises five chapters that explore the relationships between wellbeing and place in ways that speak directly to the concerns of policy-makers. In this section, place is largely positioned as the context in which wellbeing as an outcome emerges. Whilst complying with the current dominant approach to wellbeing, the studies illustrate the value of such an approach for describing problems, unpicking issues for further exploration and indicating the need for a critical but also practical approach to engaging with policy alternatives. The section opens with an extensive review of the contributions from human geography to understanding the relationships between wellbeing and place, and specifically in relation to wellbeing as health. David Conradson describes the historical trajectories of different conceptual engagements within geographies of health with the notion of wellbeing. These range from more collective welfare understandings through to the current interest in personal subjective assessments. In the second chapter, Helen Beck reports the outputs of a programme of research funded by the former English Commission for Architecture and the Built Environment (CABE) to build an evidence base on the state of urban green space and its role in enhancing health and wellbeing, especially in areas of deprivation. As such, the research programme aimed to address a knowledge gap that had been repeatedly identified by other policy agencies (for example, SDRN 2005). Beck's chapter reviews the benefits of green spaces for both local areas and residents and the extent of spatial and social variations in availability and accessibility. In reviewing the benefits of green spaces, Beck provides intimations of an understanding of wellbeing that goes beyond the personal in terms of the environmental and economic contributions, echoing uses of the concept in the literature on sustainable development. The third chapter continues with a theme of inequalities between different spatially and socially defined population groups in the availability and accessibility of resources for personal wellbeing. Mylène Riva and Sarah Curtis provide a quantitative analysis of the issues affecting wellbeing, defined as positive and self-reported health, in rural areas, drawing on secondary data from the Health Survey for England. Their detailed analysis challenges the common perception that rural areas are relatively better in terms of positive health by demonstrating the extent of diversity. In particular, they illustrate the necessity for policy makers to consider more carefully the factors which can impact on positive health in rural areas, particularly employment status and aspects of the life-course.

The next two chapters develop considerations of wellbeing at different stages in the life-course in more depth. Rose Gilroy describes the spatial and social challenges for wellbeing and ageing within an examination of the current impacts of a range of public policy measures in the United Kingdom. Action to realize the contemporary agenda for active ageing, in terms such as the adoption of new roles and identities and engagement in a range of leisure and voluntary, is dependent on facilitative local environments. The removal of both public and private provision of local amenities and services increasingly in favour of larger or cost-effective centralized outlets is just one of a number of changes in society that has negative implications for the wellbeing of older residents. Gilroy also brings to our attention the need to include consideration of how older citizens inscribe local places, the home and the neighbourhood, with meaning and express strong affective attachment to these places. Gordon Jack in his chapter takes this theme of attachment to place as his central focus and demonstrates the significance of this for children's wellbeing and the development of identity. Moreover, he identifies how, in the United Kingdom, common policy measures directed to children's wellbeing and development pay scant regard to place attachments of young people. As is often the case in literature on young people, Jack emphasizes identity and wellbeing as a dynamic process, a constant and interrelated becoming of both identity and wellness, something that Gilroy hints at in relation to older citizens. Jack argues for the value of attention to place attachments within the current climate of public spending austerity for engaging directly and meaningfully with some of the most disadvantaged children to enhance their potential wellbeing.

The next chapter explicitly juxtaposes different discourses of wellbeing and explores how the different implications for action implied therein present confusion, tensions and need for active negotiation on the part of young people. Andrea Wheeler examines initiatives in schools for more sustainable living that have been made through the design of buildings, eco-friendly measures and education for pro-environmental and sustainable behaviours. At the same time, young people are exposed constantly to messages that promise individual hedonic wellbeing through non-sustainable consumption. The importance for school environments to realize a space in which young people can actively confront, consider and negotiate conflicting social pressures is essential in order to realize both individual and collective forms of sustainable wellbeing. The final chapter in this section explores how modern culture in countries such as the United Kingdom may impact on wellbeing. Sandra Carlisle and her colleagues draw on existing literature to identify particular values that characterize modern culture, including economism, materialism, consumerism and individualism, which may impact on our experiences of personal, subjective wellbeing. The chapter demonstrates not only the significance of these values in shaping people's lives but also how the meanings and experiences of these values are situated and differ across different urban social spaces in Scotland. Carlisle and colleagues conclude this first section with a reflection on the imperative to imagine different cultural values by which to re-conceptualize an emplaced and sustainable future human wellbeing.

The chapters of the first section bring to our attention a series of concerns that policy makers tend to overlook, including the contribution of green spaces for wellbeing, the hidden diversity in rural populations, the importance of affective meanings and attachments and the different experiences for wellbeing of wider cultural values. They also expose tensions between the situated values of modern western culture that promise individual, hedonic wellbeing through modern consumer culture and the similarly situated values that urge greater attention to social connectedness, environmental responsibility and consideration for future generations. The chapters in the second section further explore these tensions in situated experiences of wellbeing. In doing so, they begin to contest the mobilization of place as a contextual backdrop, the meanings, not just the determinants, of wellbeing in different social spaces and the situated relations of power which privilege some meanings over others.

The section begins similarly to the first section with a consideration of wellbeing and green spaces. Stuart Muirhead examines the connections between environmental volunteering, green spaces and personal wellbeing. His chapter draws out contradictions in what volunteers report as the beneficial experiences of environmental volunteering for wellbeing. On the one hand, a green space that is a removal from everyday working spaces affords a therapeutic landscape of calm and retreat; on the other hand, the volunteers' tasks involve physical exertion and destruction of existing, undesirable plant life in environmental management. Muirhead argues that these apparent contradictions are reconcilable by situating wellbeing not as the hedonic elements of either emotional calm or embodied exertion, but as the eudaimonic elements of meaning and purpose. The tasks undertaken by the volunteers contribute to a wider environmental goal and enterprise that has profound ethical meanings for the volunteer, and it is through nourishing this ethical connection with place that wellbeing is produced.

Understanding personal wellbeing as embedded with ethical purpose and value has received very little direct attention in wellbeing studies and whilst the majority of chapters in this collection are drawn from settings in high income English-speaking countries, the next two chapters offer a contrast to this bias. In doing so they demonstrate how competing notions of wellbeing are both shaped by and shape the experience of particular places. Rebecca Schaaf examines the competing visions of wellbeing emerging in relation to national and global forces for modernization in Thailand. The study exposes policy tensions between reconciling global forces that promote individualized aspirations and wellbeing on the one hand and national strategies that build on and value more traditional forms of community cohesion and unity on the other. These tensions are particularly evident in the new peri-urban areas, described by Schaaf as dynamic villages in the process of a modernizing transition. Through a case study of the motivations to participate and invest in local group enterprises, Schaaf demonstrates how competing wellbeing goals are expressed and negotiated through the dynamics of a particular place. Melania Calestani similarly examines the understandings of wellbeing and place within the dynamic social spaces of urban and peri-urban Bolivia. Within a renaissance of Aymara politics, contested meanings of wellbeing, translated as living the good life, are distinguished between two geographically proximate communities in the city of El Alto in terms identity, morality and the nature of the residential neighbourhood. One community tends to define the neighbourhood and their own identities as modern and urban, aligned with the cityscape; the other, whilst dependent on the city, reaffirm their rural roots and define their neighbourhood as moral in opposition to the urban. Calestani shows how multiple meanings given to wellbeing, place and politics may emerge within culturally specific traditions in diverse mutually constituting encounters.

The next chapter returns to the United Kingdom, but resonates with Calestani's description of multiple meanings within similar social and cultural space. Karen Scott analyses the planning and implementation processes of a place-shaping intervention in an area of marked deprivation. The intervention included the purchase and demolition of a residential estate as part of local regeneration and the study demonstrates how the discursive framings of wellbeing mobilized by policy makers conflicted with local residents' perceptions of their own wellbeing. The experience led planners to reflect on their processes and to give greater consideration to who might and might not benefit in terms of wellbeing from the proposed interventions. Scott concludes by arguing for a stronger process of local participation in planning which can develop a locally-owned understanding of wellbeing. The existence of different and conflicting conceptualizations of wellbeing within the same neighbourhood and the processes through which certain conceptualizations may become dominant over others is explored again in Lorraine Gibson's chapter on contested meanings of wellbeing between Aboriginal and non-Aboriginal residents in a small country town in New South Wales, Australia. The profoundly different nature of Aboriginal cultural identity is not well understood or accommodated by the white population and, indeed, is at times treated with cynicism or scepticism. Given the greater power and dominance of white perspectives, Gibson locates a fracturing of Aboriginal cultural identity within this conflict between worldviews of how wellbeing and place are imagined.

The imagining of the connections of wellbeing and place within a critical understanding of power are also attended to, although in a very different setting, in Jo Little's chapter on the contemporary spa. The relationships between wellbeing and place are examined through the intersections of three different theoretical strands of geographical research. The chapter first explores the spa as a therapeutic landscape but then opens a more critical examination through the relations of gender, the body and identity and finally draws in approaches that offer a more trenchant critique of wider societal shifts in self-care as regulation of the body. The modern spa, as a site explicitly dedicated to personal wellbeing, affords a particularly rich site through which to interrogate the intersections between different conceptualizations of wellbeing, place and power from the body through to wider society. The final chapter goes some way towards an endeavour to bring together several of the conceptual riches in the various chapters in this volume. Ian Wight presents a theoretical exposition of how ideas of wellbeing may be incorporated into the application of integral theory, as developed by Ken Wilber. The chapter closes the volume by providing us with intriguing possibilities through which researchers may be able to undertake a complex and sophisticated engagement to explore the mutual constitution of wellbeing and place.

References

Airey, L. 2003. "Nae as nice a scheme as it used to be": Lay accounts of neighbourhood incivilities and well-being. *Health & Place*, 9, 129-37.

Albor, C. 2009. How much can money buy happiness? Is the debate over for the Easterlin Paradox? *Radical Statistics*, 98, 38-49.

Anand, S. and Sen, A.K. 1994. *Human Development Index: Methodology and Measurement*. Human Development Report Office, Occasional Papers no. 8. UNDP, New York.

Atkinson, S. 2011. Care is needed in moves to measure wellbeing. *BMJ*, 343, 1051.

Atkinson, S. and Joyce, K.E. 2011. The place and practices of wellbeing in local governance. *Environment and Planning C*, 29, 133-48.

Brereton, F., Clinch, J.P. and Ferreira, S. 2008. Happiness, geography and the environment. *Ecological Economics*, 65, 386-96.

Collins, D. and Kearns, R. 2005. Geographies of inequality: Child pedestrian injury and walking school buses in Auckland, New Zealand. *Social Science and Medicine*, 60, 61-9.

Cresswell, T. 2004. *Place: A Short Introduction*. Oxford: Blackwell.

Cronin de Chavez, A., Backett-Milburn, K., Parry, O. and Platt, S. 2005. Understanding and researching well-being: Its usage in different disciplines and potential for health research and health promotion. *Health Education Journal*, 64, 70-87.

Cummins, S. and Fagg, J. 2011. Does greener mean thinner? Associations between neighbourhood greenspace and weight status among adults in England. *International Journal of Obesity*, Advance online publication doi:10.1038/ijo.2011.195.

Davidson, J. and Milligan, C. 2004. Embodying emotion sensing space: Introducing emotional geographies. *Social and Cultural Geography*, 5, 523-32.

Deci, E.L. and Ryan, R.M. 2008. Hedonia, Eudaimonia and Well-being: An introduction. *Journal of Happiness Studies*, 9, 1-11.

DEFRA. 2005. *Securing the Future: Delivering the UK Sustainable Development Strategy*. London: DEFRA.

Diener, E., Lucas, R., Schiimmack, U. and Helliwell, J. 2009. *Well-being for Public Policy*. Oxford: Oxford University Press.

Easterlin, R.A. 1995. Will raising the incomes of all increase the happiness of all? *Journal of Economic Behaviour and Organization*, 27, 35-47.

Easterlin, R.A. 1974. Does economic growth improve the human lot, in *Nations and Households in Economic Growth: Essays in Honour of Moses Abramovitz*, edited by P.A. David and M.W. Reder. New York: Academic Press, 89-125.

Ereaut G. and Whiting R. 2008. *What do We Mean by 'Wellbeing'? And Why Might it Matter?* Linguistic Landscapes Research Report no. DCSF-RW073 for the Department for Children, Schools and Families.

Eyles, J. and Williams, A. (eds) 2008. *Sense of Place, Health and Quality of Life.* Aldershot: Ashgate.

Fleuret, S. and Atkinson, S. 2007. Wellbeing, health and geography: A critical review and research agenda. *New Zealand Geographer*, 63, 106-18.

Gallent, N. and Shaw, D. 2007. Spatial planning, area action plans and the urban-rural fringe. *Journal of Environment, Planning and Management*, 50, 617-38.

Gough, I. and McGregor, J.A. (eds). 2007. *Wellbeing in Developing Countries.* Cambridge: Cambridge University Press.

Groenewegen, P.P., van den Berg, A.E., de Vries, S., Verheij, R.A. 2006. Vitamin G: Effects of green space on health, well-being, and social safety *BMC Public Health*, 6, 149.

Hall, E. 2010. Spaces of wellbeing for people with learning disabilities. *Scottish Geographical Journal*, 126, 275-84.

Herrick, C. 2010. Challenging assumptions: Teaching, documenting, producing and negotiating 'health'. *Journal of Geography in Higher Education*, 34, 345-62.

Huppert, F.A. and Whittington, J.E. 2003. Evidence for the independence of positive and negative well-being: Implications for quality of life assessment. *British Journal of Health Psychology.* 8, 107-22.

Kahnemann, D., Diener, E. and Schwarz, N. (eds). 1999. *Well-being: The Foundations of Hedonic Psychology.* New York: Russell Sage Foundation.

Kearns, G. and Reid-Henry, S. 2009. Vital geographies: Life, luck, and the human condition. *Annals of the Association of American Geographers*, 99, 554-74.

Layard, R. 2005. *Happiness: Lessons from a New Science.* London: Allen Lane.

Mullins, D. and van Bortel, G. 2010. Neighbourhood regeneration and place leaderships: Lessons from Groningen and Birmingham. *Policy Studies*, 31, 413-28.

NEF. 2004. *A Well-being Manifesto for a Flourishing Society.* London: New Economics Foundation.

NEF, 2005. DEFRA project 3b: Sustainable development and well-being: Relationships, challenges and policy implications. Report by NEF for DEFRA, London.

NEF. 2010. *The Role of Local Government in Promoting Wellbeing.* London: NEF.

Nussbaum, M.C. 2000. *Women and Human Development: The Capabilities Approach* Cambridge: Cambridge University Press.

ONS. 2011. *Measuring National Well-being: Discussion Paper on Domains and Measures.*

Pacione, M. 2003. Urban environmental quality and human wellbeing: A social geographical perspective. *Landscape and Urban Planning*, 65, 19-30.

Panelli, R. and Tipa, G. 2009. Beyond foodscapes: Considering geographies of indigenous well-being. *Health & Place*, 15, 455-65.

Ryff, C.D. 1989. Happiness is everything, or is it? Explorations on the meaning of psychological wellbeing. *Journal of Personality and Social Psychology*, 57, 1069-81.

Ryff, C.D., Singer, B.H. and Love, G.D. 2004. Positive health: Connecting well-being with biology. *Philosophical Transactions of the Royal Society B*, 359, 1383-94.

SDRN. 2005. *Wellbeing: Concepts and Challenges*. SDRN briefing 3, London: Sustainable Development Research Network, Policy Studies Institute.

Searle, B., Smith, S.J. and Cook, N. 2009. From housing wealth to well-being? *Sociology of Health and Illness*, 31, 112-27.

Seligman, M. 2002. *Authentic Happiness*. New York: Free Press.

Seligman, M. 2011. *Flourish: A Visionary New Understanding of Happiness and Well-being*. New York: Free Press.

Shneekloth, L. and Shibley, R. 1994. *Placemaking: The Art and Practice of Building Community*. New York: Wiley.

Singh-Manoux, A., Marmot, M.G. and Adler, N.E. 2005. Does Subjective Social Status Predict Health and Change in Health Status Better Than Objective Status? *Psychosomatic Medicine*, 67, 855-61.

Smith, D.M. 2000. *Moral Geographies: Ethics in a World of Difference*. Edinburgh: Edinburgh University Press.

Smith, G. and King, M. 2008. Naturism and sexuality: Broadening our approach to sexual wellbeing. *Health & Place*, 15, 439-46.

Sointu, E. 2005. The rise of an ideal: Tracing changing discourses of wellbeing. *Sociological Review*, 53, 255-74.

Steuer, N. and Marks, N. 2008. *Local Wellbeing: Can we Measure it?* London: The Young Foundation and NEF.

Stevenson, B. and Wolfers, J. 2008. *Economic Growth and Subjective Well-being: Reassessing the Easterlin Paradox*. National Bureau for Economic Research Working Paper No. 14282. NBER, Massachussetts.

Stiglitz J., Sen A. and Fitoussi J. 2009. *Report by the Commission on the Measurement of Economic Performance and Social Progress* [Online]. Available at: www.stiglitz-sen-fitoussi.fr [accessed: 29 September 2009].

Wellbeing and Poverty Pathways. 2011. *An Integrated Model for Assessing Wellbeing*. Wellbeing and Poverty Pathways Briefing No. 1. Brunel and Bath Universities.

WHO, 1948. Constitution of the World Health Organization. WHO, Geneva.

Wilkinson, R. and Pickett, K. 2009. *The Spirit Level*. London: Penguin.

Wilkinson, R. and Marmot, M. (eds) 2003. *Social Determinants of Health: The Solid Facts*. Geneva: WHO.

Chapter 2

Wellbeing: Reflections on Geographical Engagements

David Conradson

Wellbeing has attracted significant academic and political attention in recent years. Researchers in psychology (Diener et al. 1998, Kahneman et al. 1999, Huppert 2009), economics (Blanchflower and Oswald 2004, Layard 2005), public health (Carlisle and Hanlon 2008, Crawshaw 2008) and development studies (Sen 1992, Camfield et al. 2009, 2010) have sought to conceptualize, measure and explain variations in wellbeing between individuals and groups. In the political arena, several western governments have commissioned reports on national wellbeing in recent years, including France, Canada and Britain (Dolan et al. 2011, Stiglitz et al. 2009, Waldron 2010). Many not-for-profit and third sector organizations are also turning their attention to the idea of wellbeing, with activities that include public initiatives to promote wellbeing (New Economics Foundation 2004, 2008) as well as new ways to measure and visualize wellbeing data (New Economics Foundation 2010).

This chapter examines geographical engagements with wellbeing. I focus on two particular streams of work: firstly, a welfare-oriented mobilization of wellbeing associated with social geography and, secondly, wellbeing scholarship that has emerged from within health geography. The chapter has four main parts. I begin by discussing some of the conceptual and methodological issues that concern social research on wellbeing. In the second part of the chapter, I review a selection of welfare-oriented geographical work on wellbeing, including the development of territorial social indicators during the 1970s, work on quality of life and life satisfaction liveability, and more recent investigations into the geographies of happiness and cultural constructions of wellbeing. A good deal of this welfare-oriented research has employed quantitative methods and, in its early years, it tended to rely primarily on objective measures of wellbeing. More recently, however, there has been a shift towards subjective assessments of wellbeing and an increased use of qualitative research methods. In the third part of the chapter, I consider work on wellbeing that has emerged from within health geography, including studies of therapeutic landscapes, green space, 'healthy cities' (e.g. urban environments which promote active transport and minimize the risk of obesity), and the emotional dimensions of place attachment. This research is part of an ongoing movement beyond medical geography's early focus on illness and disease towards an interest in how places facilitate health (Kearns 1993, Gesler and Kearns 2002, Kearns and Moon 2002). I conclude with some observations about geographical perspectives on wellbeing.

Conceptualizing Wellbeing

Even a cursory examination of the interdisciplinary research literature on wellbeing reveals a diversity of conceptualizations and methods. Economists approach the subject rather differently than anthropologists, for example, whilst public health researchers raise different questions again (Carlisle and Hanlon 2008). The variety of definitions and understandings in this field can present challenges for governmental and non-governmental organizations as they seek to develop wellbeing policies and practices (Atkinson and Joyce 2011). Despite this range of views, there are points of shared understanding in wellbeing research. Wellbeing is most commonly understood as a holistic conception of positive human functioning, for instance, extending beyond a physiological or biomedical notion of health to encompass the emotional, social and, in some cases, spiritual dimensions of what it means to be human. Wellbeing is also an explicitly positive concept, in that it describes the presence of positive qualities and experiences rather than simply the absence of illness and disease.

A distinction is often drawn between 'hedonic' and 'eudaimonic' conceptions of wellbeing (Waterman 1990). Hedonic conceptions understand wellbeing in terms of the experience of pleasure or happiness. While pleasure and happiness may be associated with a well-functioning human organism, however, this is not always the case. In addition, it is possible to experience these emotional states whilst participating in suboptimal, dysfunctional and even injurious patterns of behaviour. This observation suggests that pleasure is not, in and of itself, a good or sufficient indicator of wellbeing. Eudaimonic conceptions of wellbeing, which emphasize human flourishing and life satisfaction over time, go some way towards addressing these limitations of the hedonic view. From a eudaimonic perspective, it is possible for a person to experience hardship and suffering but to nevertheless report high levels of subjective wellbeing. This unexpected situation may be possible because the individual is able to view her hardship and difficulties within a wider framework or interpretative horizon.

A further distinction can be made between 'objective' (externally acquired) and 'subjective' (self-reported) measures of wellbeing. Objective wellbeing measures may be derived from published health and socio-economic data for a population of interest (e.g. of educational attainment), or by taking direct measurements from individuals (e.g. of Body Mass Index (BMI), blood pressure and stress-related cortisol). While early wellbeing research relied largely upon population-level objective indicators, in recent years there has been a turn towards subjective self-report measures, used either in combination with objective data or on their own (Diener et al. 1998, 1999, Dolan et al. 2011). There is now a considerable amount of evidence demonstrating the validity of subjective wellbeing measures (Veenhoven

1996, Diener et al. 1999), including studies demonstrating their correlation with 'objective physiological and medical criteria' (Kahneman and Kruger, 2006: 7).

Subjective wellbeing is typically understood as having two main components: an affective or emotional element, which is about a person's felt sense of their current lived experience, and a cognitive element, which entails evaluations of that life, perhaps referenced against local norms and with a degree of retrospective review. As Diener et al. (1998: 34) explain, 'subjective wellbeing is a person's evaluation of his or her life. This valuation can be in terms of cognitive states such as satisfaction with one's marriage, work and life, and it can be in terms of ongoing affect (i.e., the presence of positive emotions and moods, and the absence of unpleasant affect)'. Some subjective measures of wellbeing focus primarily on cognitive judgements of life satisfaction (e.g. the World Values Survey asks its respondents the following question: 'All things considered, how satisfied are you with your life these days?), whilst other measures concentrate on the affective or emotional dimensions of wellbeing.

Wellbeing researchers have sought to understand what makes people happy, with investigations that encompass genetic and behavioural factors as well as social and environmental determinants. The underlying questions in this research are not, of course, new and Aristotle's work on human flourishing is an important philosophical touchstone in many cases (Vernon 2005). For contemporary wellbeing research, however, Easterlin's (1974) work on the 'paradox of affluence' might be regarded as a foundational study. Easterlin examined life satisfaction and GDP per capita for the American population during the post-war period. The results were striking, for while GDP per capita increased significantly through the post-war decades, levels of self-reported life satisfaction flat-lined. In short, despite remarkable increases in material affluence, the American people appeared little happier on average. Subsequent investigations revealed that increases in income may have a positive effect on subjective wellbeing when a person is very poor, but that this positive effect quickly attenuates once the individual moves beyond the level of material subsistence (Layard 2005, Eckersley 2009). Easterlin's (1974) overall thesis has been corroborated in a number of other studies (e.g. Blanchflower and Oswald 2004), which essentially show that, once a baseline level of material needs are met, economic growth does not translate straightforwardly into increased subjective wellbeing. This work has prompted the development of alternative, non-GDP based measures of economic and social progress. In Australia and Britain, there are now initiatives to develop national measures of subjective wellbeing, for instance, whilst the nation of Bhutan has developed a Gross National Happiness index (New Economics Foundation 2010).

Welfare Oriented Geographies of Wellbeing

Having briefly addressed some of the key ideas in the interdisciplinary field of wellbeing research, we can now consider how geographers have engaged with the issue. While wellbeing has been a recent focus of interest amongst geographers (Fleuret and Atkinson 2007, Kearns and Andrews 2010), geographical engagement with the topic in fact extends back to the welfare geographies of the 1970s. At this time, a number of geographers sought to develop multivariate indicators of wellbeing, quality of life and 'level of living' for particular localities (Smith 1973, Knox 1974a, 1974b, Knox 1975). This scholarship built upon the social indicator tradition in the United States, which had produced national measures of social progress by compiling relevant data at regional, metropolitan and neighbourhood scales (Wilson 1969, Andrews and Withey 1976). The geographers' aim was to construct 'territorial social indicators', thereby overcoming the lack of spatial differentiation that characterized national-level economic and social statistics.

In conceptual terms, Smith (1973) argued that an individual's wellbeing in a given location could be understood as the relationship between their actual living conditions – their observed quality of life – and their expected quality of life or what they needed. A measure of *social* wellbeing could then be derived from the average difference between expected and observed values for individuals within a given group or territorial area, with the addition of an interaction term (to account for the group dynamics which might enhance or detract from the average level of wellbeing in a given population) (Smith 1973: 74). This conceptualization was difficult to operationalize, however, as the wellbeing data available at the time consisted primarily of objective indicators such as employment status, educational attainment and income. In the absence of commissioned survey work, there was little understanding of how variations in these factors were subjectively experienced by individuals and communities, let alone what people expected or needed. As Smith (1973) acknowledged, the territorial social indicators tradition thus effectively dealt with surrogate rather that direct measures of subjective wellbeing (in the sense of how a person evaluated and felt about their life circumstances).

This limitation notwithstanding, Smith used a range of objective measures to map variations in social wellbeing across the United States (US). Drawing on existing research, he identified seven broad domains of human life that were understood to contribute significantly to human wellbeing: (1) income, wealth and employment; (2) the living environment; (3) health; (4) education; (5) social order; (6) social belonging; and (7) recreation and leisure. The terrain covered by these domains was extensive, encompassing the economy (e.g. employment levels in local labour markets), the physical and residential environment, social dynamics, and a sense of belonging or connection to the collective. For a given spatial unit of analysis – such as a city or region – the challenge was to identify a set of variables which (a) related meaningfully to these wellbeing domains and (b) for which secondary data was publicly available (as commissioned survey work would seldom provide the desired spatial coverage and was in any case expensive). As an example of this process of operationalization, Smith (1973) measured *health* at the US state level through a set of variables relating to *physical health* (e.g. proportion of households with poor diets, infant mortality rates, hospital expenditure per patient day), *mental health* (e.g. in- and out-patient usage of mental health services) and *health service accessibility* (e.g. numbers of hospital beds, physicians and dentists per one thousand people). This approach built closely on the multivariate social indicators tradition, something which continues to inform contemporary work on deprivation and wellbeing.

Drawing on 1960s data for over 40 separate indicators, Smith (1973) found significant geographical inequalities in social wellbeing within the US. His analysis highlighted the southern states of Alabama, Arkansas, Georgia, Louisiana, Mississippi, North Carolina, South Carolina, and Tennessee as places characterized by low social wellbeing. In contrast, the best states on most wellbeing measures were located in the west (California, Oregon, Washington), the midwest (e.g. Wisconsin, Iowa, Minnesota), and the Northeast (e.g. Massachusetts). This specification and the related visual representations of the geography of social wellbeing in the US were an important research contribution, and more novel than contemporary readers might appreciate. Smith understood that description was only part of the research task, however, and also undertook some preliminary work to ascertain the processes underlying the observed inequalities. Using a principal components analysis, the most significant predictors of state-level social wellbeing were found to be high income (presumably because it afforded access to good housing, education and healthcare, as well as to enabling forms of social capital), low levels of social disorganization, and low levels of social pathology (e.g. crime and drug use). Given the cross-sectional nature of the dataset, Smith was not able to explore whether these associations were operating as causal pathways, or how they might be doing so, but his work was significant as an early geographical exploration of the associations between population wellbeing and a range of demographic, behavioural and contextual factors. In addition to his state-level analysis, Smith examined variations in quality of life between a selection of US cities, as well as *intra*-urban variations in wellbeing within the city of Tampa, Florida.

In contrast to what one might assume from characterizations of quantitative geography during the 1960s and 1970s as largely apolitical, the early work on territorial social indicators sought to reveal under-appreciated inequalities and social injustices. As Smith explained:

> [t]erritorial social indicators ... portray an important yet often neglected dimension of the social system. They reveal the extent to which groups of people, defined by area of residence, have different experiences with respect to conditions that have a bearing on the wellbeing of society and the quality of individual life. They show that levels of income (broadly defined), environmental quality, health, education, social disorganisation, alienation, and participation are subject to consideration variation between territories. There are extreme inequalities at all spatial levels. (1973: 139-40)

In suggesting that socio-spatial inequalities were neglected by researchers, Smith's work is typical of the broader welfare geography of the period.[1] Much of this scholarship sought to document the nature and inequalities of people's circumstances across different places and regions. The implicit emphasis on the social and structural determinants of wellbeing would become a major focus in the radical and Marxist geography that developed through the 1970s (e.g. Peet 1975, Harvey 1972, 1973, Castree et al. 2010). To varying degrees, this desire to identify the processes underlying observed socio-spatial inequalities has been a signature characteristic of geographical work on wellbeing since that time (e.g. Dorling 2011).

The ethos and methods of this territorial social indicators work were taken up a number of other geographers. In New Zealand, for example, McCracken (1977, 1983) drew upon a range of published health, social and population data to compile a multi-dimensional wellbeing indicator for census area units across the country. This indicator encompassed six main domains (economic, health, social order, education, social belonging, and recreation and leisure). Despite the moderating influence of a redistributive post-war welfare state, the findings demonstrated that:

> ... New Zealand clearly has a geography of social wellbeing. While the country may be fortunate in that its extremes of territorial social inequality are generally not as severe as those found in most other western world nations, they are nonetheless sufficiently substantial to warrant public and governmental concern. (McCracken 1977: 31)

This work was updated a decade later (Kane and Wards 1989), in a study that revealed similar patterns of affluence and disadvantage across the country. The approaches employed in these and other territorial social indicator studies formed an important basis for later geographical investigations of the spatial patterning of poverty, deprivation and inequality (e.g. Pacione 1982, 1986, Dorling 1995, Philo 1995, Glasmeier 2005).

From the late 1970s onwards, the notions of *quality of life* and *life satisfaction* emerged as another way in which geographers and others conceptualized wellbeing (Kuz 1978, Kearsley 1982, Pacione 1980). Most of this work was based on objective indicators, using data relating to factors such as educational achievement, environmental quality and access to services. With respect to the physical and built environment, a number of studies found negative associations between life satisfaction and air pollution (e.g. Welsch 2006, Smyth et al. 2008, Luechinger 2009), urban noise, traffic congestion (Stutzer and Frey 2008) and population density (Cramer et al. 2004; Walton et al. 2008). Other quality of life studies focused more on the social environment, as in Pacione's (2003) work on Glasgow, where he developed an intra-urban index of multiple deprivation

---

1   Smith's statement might also be said to anticipate his later work on the geographies of justice (e.g. Smith 1994).

by compiling data on male unemployment, council housing, single parents, and household occupancy in particular locations. As with the early work on territorial social indicators, these quality of life studies tended to interpret the absence of negative indicators as the presence of positive quality of life. This approach is evident in Pacione's (1980, 1986) research, as well as in studies of social wellbeing from Australia (Sorensen and Weinand 1991, Walmsley and Weinand 1997) and India (Kulkarni 1990).

In contrast to this objectively focused work, Morrison (2007, 2010) drew on biennial surveys to examine subjective evaluations of quality of life across New Zealand 12 largest cities. The surveys questioned 500 respondents in each city regarding their Happiness, Satisfaction with Life, and perceived Quality of Life. The resulting data enabled Morrison to work directly with a positive formulation of quality of life, rather than having to infer its presence from the absence of negative factors. As well as identifying differences in subjective wellbeing between cities, the research sought to ascertain how much of the observed differences could be explained by the socio-economic characteristics of local populations and whether 'place effects' were also in operation.

After controlling for compositional factors such as health and marital status, Morrrison (2010) found significant place effects for self-reported happiness across New Zealand's 12 main cities, with a smaller but nonetheless significant effect for life satisfaction and quality of life. Auckland, the country's largest urban centre by some margin, scored lowest on both the Happiness and Satisfaction with Life measures (the affective and cognitive dimensions of subjective wellbeing respectively). However self-reported Quality of Life amongst Auckland respondents was better than what the city's population density and intra-urban commuting might suggest. Morrison suggested that Auckland's relatively favourable evaluation on this measure might reflect the work and income possibilities associated with a dynamic labour market, or relate to environmental factors such as the favourable climate, neither of which were considered in the survey. In general, however, the geographic distribution of happiness in New Zealand does not mirror the national geography of wealth (which is concentrated in Auckland). With reference to Easterlin's (1974) classic work, Morrison (2010) thus described his findings as 'the localisation of the paradox of affluence'.

The recent emergence of geographical work on happiness adds a further dimension to the stream of welfare oriented geographical work on wellbeing. As Layard's (2005) popular volume *Happiness: Lessons from a New Science* demonstrates, the dynamics of human happiness is now a major research area within psychology, economics, genetics and neuroscience (Frey 2008, Brockman and Delhey 2010). There are a number of key questions in this work. For example, how much of a person's happiness is determined by their circumstances, genetics, and voluntary activities? What is the effect of negative events on human happiness, and why do individuals recover or re-equilibrate from the same events at such different rates? How might we understand those cases where people's happiness appears to collapse in response to stressor events, and what forms of relational or pharmacological interventions might best moderate or offset their associated distress? Some of this research deploys happiness as a synonym for subjective wellbeing, and addresses both its cognitive and affective dimensions. But it has nevertheless provided another avenue for geographical research.

In this regard, Ballas and Dorling (2007) drew on the British Household Panel Survey to estimate the impacts upon people's happiness of major life events such as starting/ending a relationship, gaining/losing employment, and the death or illness of a parent. In a subsequent study, Ballas and Tranmer (2012: 70) employed a multi-level modelling approach to 'assess the nature and extent of variations in happiness and wellbeing to determine the relative importance of the area (district, region), household, and individual characteristics on these outcomes'. By controlling for variations at the individual and household levels, the authors were able to determine whether there were any statistically significant geographical differences within Britain in terms of self-reported happiness. Although there were apparent geographical differences, the analysis suggested that compositional (i.e. demographic) rather than contextual (i.e. place) factors were in fact the primary determinants of variations in self-reported happiness across the country. The authors did note, however, that 'length of time in current address ... had a positive and significant effect on subjective wellbeing'. This effect might relate to the time needed to develop localized social networks and connectedness, resources that are known to confer resilience to external stressors and promote mental health (Layard 2005). Geographically oriented work on happiness is also being undertaken outside of geography, including studies of the relationship between income and happiness for a sample of 94 countries (Stanca 2010) and use of the International Social Survey Programme (2002) to compare self-reported happiness between 35 countries (Blanchflower and Oswald 2005).

As is evident from the discussion so far, many of these geographical studies of wellbeing, quality of life and happiness employ quantitative data and methods. Over the last decade or so, however, a number of more qualitative social geographical studies of wellbeing have emerged. One theme in this work has been to investigate how local communities and cultural groups understand wellbeing on their own terms, rather than assuming that externally imposed objective measures are inevitably meaningful or relevant. Panelli and Tipa (2007) model such an approach in their discussion of Māori wellbeing in Aotearoa New Zealand. Māori understand wellbeing as something derived through social, ecological and spiritual relationships. Wellbeing for Māori is closely influenced by connections with the land, to other people, and to something bigger than oneself. As the authors note, such a perspective differs significantly from something like Ryff's (1989) six-factor psychological model, which conceives of wellbeing in terms of '... autonomy, environmental mastery, personal growth, positive relations with others, purpose in life, and self-acceptance' (Panelli and Tipa 2007: 446). Ryff's work appears to assume a relatively autonomous individual, somewhat independent of connection to a wider community or to the earth. Panelli and Tipa's (2007, 2009) analysis shows that psychological models developed in particular cultural settings are far from universally applicable; wellbeing is in fact always experienced in relation to particular places and environments. This perspective resonates with other studies of indigenous health and wellbeing knowledges (e.g. Adelson 2000, Ingersoll-Dayton et al. 2004, Holmes et al. 2002, Izquierdo 2005), whilst there are also links to participatory approaches for understanding culturally specific constructions of wellbeing (e.g. Crivello et al. 2009). A similar approach was adopted by Scott (2012) in her work on local constructions of wellbeing in the northeast of England.

Health Geographies and Wellbeing

If welfare oriented analyses are one significant stream of geographical scholarship on wellbeing, then health geography is the other. The context for such work is the move beyond medical geography, with its focus on the distribution of illness, disease and inequalities in access to health services, towards a more broadly constituted interest in the relation between health and place (Kearns 1993, Kearns and Gesler 1998, Kearns and Moon 2002). In fact, if we understand 'health geography' as a broadening of the field of 'medical geography', then the 'geography of wellbeing' might be understood as a further expansion of the field's focus and content. Each of these expansions should ideally be inclusive of previous scholarship, such that contemporary health geographers can continue to engage with the clinical and organizational spaces of medical and other forms of healthcare practice (Andrews and Evans 2008).

Four areas of health geographic scholarship on wellbeing are of particular interest here. The first concerns therapeutic landscapes, a body of work which emerged in the wake of Gesler's (1992) seminal article about the therapeutic capacities of the built, social and symbolic dimensions of places such as Epidaurous (a hot spring in Greece culturally recognized as a 'healing place'). Through a diverse set of case-studies that include alpine settings, forests, beaches, drop-in centres, summer camps, retreat centres and sacred places (Williams 1999, 2008), the therapeutic landscapes literature has highlighted the positive emotional and bodily effects of spending time within particular kinds of 'natural' environments. Most of this research has employed qualitative methods, in an effort to stay close to the felt experience and narratives of people spending time in such places. The rich accounts which emerge differ from the more quantitative and quasi-experimental psychological research in the related field of restorative environments (Hartig et al. 2003, Kaplan 1995). As well as building upon Gesler's (1992) seminal work, some studies of therapeutic landscapes actively draw upon concepts, methodologies and writing strategies from the fields of cultural and emotional geographies (Anderson and Smith 2001, Davidson et al. 2005; Conradson 2005, Foley 2010). There are also connections to social geographic work on supportive environments more generally, as part of work on geographies of care (Conradson 2003, DeVerteuil and Wilton 2010, Parr and Philo 2003) and voluntary welfare provision (Milligan and Conradson 2006, Wilton and DeVerteuil 2006).

A second contemporary area of health geographic work on wellbeing concerns 'greenspace' (Mitchell et al. 2011, Mitchell and Popham 2007, de Vries et al. 2003). Working in the tradition of social epidemiology, but with a health rather than ill-health focus, this research has examined the potential of parks and other forms of urban green spaces to promote health. There is some evidence that proximity to urban greenspace does have a salutogenic effect, after controlling for so-called 'selection effects' (e.g. people living near parks are likely to be somewhat wealthier, and wealthy people are generally healthier than their low-income counterparts) (Maas et al. 2006, Richardson et al. 2010). Groenewegen et al. (2006) refer to this greenspace effect as 'vitamin G', with the suggestion being that developing new urban parks and enhancing access to the existing ones may be a more cost-effective public health intervention than spending large amounts of public money addressing the obesity, coronary heart disease and diabetes associated with sedentary lifestyles. As yet, however, it is not exactly clear what it is about proximity to urban greenspace that promotes health. There is some evidence that access to greenspace may promote increased physical activity, such as walking, running, and team sports, and that this in turn improves cardiovascular health and protects against obesity (Hillsdon et al. 2006, Maas et al. 2008). There may also be some psychologically calming effect as a result of being able to look onto and, at a city scale, be located within pleasant green environments (Ulrich 1984). Mitchell et al. (2011) investigated whether the physical size of the green space influences the health promoting effect, and found some evidence to suggest that larger greenspaces might generate larger health effects than their smaller counterparts. The 'quality' of the greenspace, in terms of how clean, safe and well lit after dark it is perceived to be, is also a significant determinant of how well it is used. A final point is that access to greenspace is typically uneven across western cities, and so researchers have sought to identify whether social gradients in access to greenspace might exacerbate or offset existing forms of health inequality.

A third body of work relating to health geography and wellbeing explores the issue of 'healthy cities'. The basic concern here is that the low density urban forms which are so prevalent in North America and Australasia tend to encourage reliance upon cars for travel, reducing a person's daily exercise levels and thus their bodily energy output, in turn making it easier to gain weight and develop associated health problems. There is an opportunity, then, for public health specialists, urban planners, geographers and other social scientists to come together to share expertise and knowledge around designing healthy cities (PHAC 2010). Desirable improvements in this regard include the promotion of active transport (i.e. cycling, walking, skating across the city, the uptake of which often depends in part upon their perceived safety); the introduction of planning regulations to restrict localized access to fast-food and tobacco-retail outlets, especially in the vicinity of schools (Day and Pearce 2011); the introduction of clean-air heating technologies (e.g. gas fires or electrical heat pumps) so as to reduce atmospheric particulates, in turn perhaps reducing respiratory problems such as asthma and bronchitis amongst the population; and the provision of adequately heated and well insulated homes, particularly for younger and older generations, so as to minimize diurnal temperature fluctuations within domestic environments. For geographers working on these kinds of topics, there are significant opportunities for collaborations with public health researchers, urban planners, architects and other social scientists (see PHAC 2010).

A fourth area of work in health geography emphasizes the significance of emotional and imagined connections to place as an important dimension of wellbeing. In Wiles et al.'s (2009) study of older people's experiences of ageing in place, the participants were receiving support services whilst living in their own homes, as opposed to transitioning to a residential care facility. This enabled them to maintain a felt and embodied connection with a familiar environment, as well as supporting localized forms of belonging and social connection. As older people become more frail and unable to travel independently, however, the risk of social isolation associated with staying in one's own home grows, and with it the potential for loneliness and depression. Garvin and Nykiforuk's (2011) work on people growing older in their own homes in suburban Alberta, Canada, highlights just this point, particularly as access to local shopping and other facilities were typically dependent upon use of a private car. Whether researching older people or some other social or cultural group, a number of geographers working on the connections between wellbeing and place attachment have drawn upon contemporary work in emotional geographies (Anderson and Smith 2001, Davidson et al. 2005). This work invites researchers to give expression to the range and depth of feelings present amongst research participants, rather than exclude these as part of a misplaced understanding of objectivity.

Discussion

As this review indicates, geographical engagements with wellbeing have taken a variety of different forms. From the development of territorial social indicators in the 1970s through to present day theorizations of therapeutic landscapes, geographers have approached wellbeing using a number of different conceptual frameworks and methodological strategies. By way of discussion, three observations can be made of this work.

The first point concerns scales and spaces of analysis. Within the welfare oriented and quantitative health geographical work on wellbeing, much of the analysis has focused on the national, regional and city scales. Researchers have mapped variations in social wellbeing within and between these territories, seeking to identify the contextual and compositional determinants of observed patterns (Duncan et al. 1998). Over time, the methods and techniques employed in this research have become more sophisticated, informed by developments in social epidemiology (Berkman and Kawachi 2000) and the evolution of multilevel modelling techniques (Rice and Leyland 1996, Snidjers and Bosker 1999). In addition to these relatively extensive analyses, health and social geographers have investigated local spaces which foster wellbeing (Fleuret and Atkinson 2007). At the neighbourhood level, research on obesogenic environments (Day and Pearce 2011), green space (Maas et al. 2006) and healthy cities (PHAC 2010) is illustrative of this work. The health effects of specific landscapes and organizational environments has also been examined, including studies of therapeutic landscapes (e.g. Conradson 2005, Foley 2010, Williams 2008) and spaces of care and support (e.g. Conradson 2003, Hall 2007, Wilton and DeVerteuil 2006).

A second characteristic of geographical scholarship on wellbeing is its consistent attention to socio-ecological context. As noted in the chapter introduction, there are currently active programmes of wellbeing research in disciplines as diverse as anthropology, economics, public health and neuroscience. Geography's distinctive contribution to this broader field of endeavour is to further understanding of the relationship between socio-ecological context, however conceived, and wellbeing. In comparison, mainstream psychological research on wellbeing often pays relatively little attention to social context beyond the immediate family. Large groups of participants may thus be dealt with as sets of individuals, without consideration of the interactions amongst them or the social structures that might lead to group effects. There are, of course, social sciences keenly interested in contextual analyses of worldly phenomena; sociology and anthropology are two obvious examples. However, the considerations of context in these disciplines tend, on balance, to focus on the social domain, foregrounding human forms of life and sometimes bracketing out the non-human, the environmental, and the technological. Geographers, on the other hand, tend to conceptualize places as localized arrangements of humans, non-humans, material and technological objects, all of which are open to revision and ongoing development.

For geographers interested in wellbeing, the question which then arises is how does the position of humans within particular socio-natural-technical configurations influence their health and wellbeing? At one end of a methodological and conceptual spectrum, this question has been framed and explored using quantitative data and experimental designs, as for example in the studies of urban green space (Mitchell and Popham 2007). At another end, students have investigated these issues using qualitative and more ethnographic methods (e.g. Wiles et al. 2009). Each of these methodological strategies, along with the possibilities between them, has the capacity to deepen our understanding of the relationship between humans, their environments and wellbeing.

A third point is that geographical engagements with wellbeing sit within a broader disciplinary tradition of critical enquiry and action. Geography has long been concerned not only to describe the world but also to consider how we might participate in changing it (Castree et al. 2010). This critical impulse is evident in the early work on territorial social indicators (Smith 1973, Knox 1975) and continues to be present in contemporary geographical engagements with wellbeing, including studies of healthy cities and urban greenspace. In much of the mainstream economic and psychological literature on wellbeing, however, there seems to be little discussion of broader social and environmental determinants of health. A good deal of psychological research refrains from exploring how individual distress and difficulty might reflect broader social and cultural factors. Unlike public health, where the social determinants of health are explicitly acknowledged (Carlisle and Hanlon 2008; Carlisle et al. 2009), many of the wellbeing interventions suggested by psychologists therefore focus on what Watzlawick et al. (1974) described as first order change (seeking to alter individual behaviours) rather than second order change (adjusting the systems and structures which contribute to individual difficulty and distress). Within economic research on wellbeing and happiness, there is also often a curious silence about the limitations and problems inherent in the capitalist economic system (e.g. Helliwell and Putnam 2004). And yet there is a strong tradition within geography of critically evaluating contemporary economic arrangements and also, to some extent, of political engagement that seeks to improve the situation of the many rather than the few (Castree et al. 2010). The work of Gibson-Graham (2006) on post-capitalist community economies is a fine example of such critical geographical scholarship, and might productively be brought into conversation with some contemporary geographical work on wellbeing. If geographers interested in wellbeing are able to draw insight from critical scholarship of this kind, we will be better placed to make a distinctive contribution to the broader interdisciplinary field of wellbeing research. As Kearns and Collins (2010: 27) note:

> The ongoing challenge for health geographers is to rethink the issue of wellbeing by contextualising it into both personal and population-based experiences of place, whilst holding firm to traditional concerns for equity and social justice.

Conclusion

Over the last 40 years, geographers have analysed wellbeing at scales ranging from the local to the national. This work has helped to reveal how aspects of the social, natural and built environment positively influence human health and wellbeing. As a consequence, we now have a better understanding of the characteristics of health-giving environments – including urban green spaces, 'healthy cities' and socially connected communities – and the mechanisms which underlie their salutogenic effects. In exploring the socio-ecological determinants of health, this geographical research complements those psychological and medical studies which emphasize individual determinants of wellbeing (including behaviours such as regular exercise, good diet and recreation). The geographer's interest in spatial variation also adds value to national level studies of wellbeing, highlighting the regions and places where people are doing well and less well. A final point is that geographical research on wellbeing is able to draw on a longstanding disciplinary tradition of critical analysis, directed in particular to the evaluation of existing social and economic arrangements. These critiques call us not to stop at analysis, but to develop recommendations for policies and practices that will actively facilitate social wellbeing.

References

Adelson, N. 2000. *"Being Alive Well": Health and the Politics of Cree Well-being.* Toronto: Toronto University Press.

Anderson, K. and Smith, S.J. 2001. Editorial: Emotional geographies. *Transactions of the Institute of British Geographers*, 26(1), 7-10.

Andrews, F.M. and Withey, S.B. 1976. *Social Indicators of Well-Being: Americans' Perception of Life Quality.* New York: Plenum Press.

Andrews, G. and Evans, J. 2008. Reproducing health care: Towards geographies in health care work. *Progress in Human Geography*, 32, 759-80.

Atkinson, S. and Joyce, K.E. 2011. The place and practices of wellbeing in local governance. *Environment and Planning C*, 29, 133-48.

Ballas, D. and Dorling, D. 2007. Measuring the impact of major life events upon happiness. *International Journal of Epidemiology*, 36, 1244-52.

Ballas, D. and Tranmer, M. 2012. Happy People or Happy Places? A Multilevel Modelling Approach to the Analysis of Happiness and Well-Being. *International Regional Science Review*, 35(1), 70-102.

Berkman, L.F. and Kawachi, I. 2000. *Social Epidemiology.* New York: Oxford University Press.

Blanchflower, D.G. and Oswald, A.J. 2004. Well-being over time in Britain and the USA. *Journal of Public Economics*, 88, 1359-86.

Blanchflower, D.G. and Oswald, A.J. 2005. Happiness and the Human Development Index: The Paradox of Australia. *The Australian Economic Forum*, 38(3), 307-18.

Brockmann, H. and Delhey, J. 2010. Introduction: The Dynamics of Happiness and the Dynamics of Happiness Research. *Social Indicators Research*, 97, 1-5.

Camfield, L., Crivello, G. and Woodhead, M. 2009. Wellbeing research in developing countries: Reviewing the role of qualitiative methods. *Social Indicators Research*, 90, 5-31.

Camfield, L., Streuli, N. and Woodhead, M. 2010. Children's well-being in developing countries: A conceptual and methodological review. *European Journal of Development Research*, 22(3), 398-416.

Carlisle, S. and Hanlon, P. 2008. 'Well-being' as a focus for public health? A critique and defence. *Critical Public Health*, 18(3), 263-70.

Carlisle, S., Henderson, G. and Hanlon, P.W. 2009. Wellbeing: A collateral casualty of modernity? *Social Science and Medicine*, 69, 1556-60.

Castree, N., Chatterton, P., Heynen, N., Larner, W. and Wright. M.W. (eds) 2010. *The Point Is To Change It: Geographies of Hope and Survival in an Age of Crisis.* Oxford: Wiley-Blackwell.

Conradson, D. 2003. Spaces of care in the city: The place of a community drop-in centre. *Social and Cultural Geography*, 4(4), 507-25.

Conradson, D. 2005. Landscape, care and the relational self: Therapeutic encounters in rural England. *Health and Place*, 11, 337-48.

Cramer, V., Torgersen, S. and Kringlen, E. 2004. Quality of life in a city: The effect of population density. *Social Indicators Research*, 69, 386-96.

Crawshaw, P. 2008. Whither wellbeing for public health? *Critical Public Health*, 18(3), 259-61.

Crivello, G., Camfield, L. and Woodhead, M. 2009. How can children tell us about wellbeing? Exploring the potential of participatory research approaches within Young Lives. *Social Indicators Research*, 90, 51-7.

Davidson, J., Bondi, L. and Smith, M. (eds) 2005. *Emotional Geographies.* Aldershot: Ashgate.

Day, P. and Pearce, J. 2011. Obesity-Promoting Food Environments and the Spatial Clustering of Food Outlets Around Schools, *American Journal of Preventive Medicine*, 40(2), 113-21.

de Vries, S., Verheij, R.A., Groenewegen, P.P. and Spreeuwenberg, P. 2003. Natural environments – healthy environments? An exploratory analysis of the relationship between greenspace and health. *Environment and Planning A*, 35(10), 1717-31.

Deverteuil, G. and Wilton, R. 2010. Spaces of abeyance, care and survival: The addiction treatment system as a site of 'regulatory richness. *Political Geography*, 28(8), 463-72.

Diener, E., Eunkook, M.S, Lucas, R.E. and Smith, H.L. 1999. Subjective well-being: Three decades of progress. *Psychological Bulletin*, 125, 276-303.

Diener, E., Sapyta, J. and Eunkook, S. 1998. Subjective well-being is essential to well-being. *Psychological Inquiry*, 9(1), 33-7.

Dolan, P., Layard, R. and Metcalfe, R. 2011. *Measuring Subjective Well-being for Public Policy*. London: UK Office for National Statistics.

Dorling, D. 1995. *A New Social Atlas of Britain.* London: Wiley & Sons.

Dorling, D. 2011. *Injustice: Why Social Inequality Persists*. Bristol: Policy Press.

Duncan, C., Jones, K. and Moon, G. 1998. Context, composition and heterogeneity. *Social Science and Medicine*, 46, 97-117.

Easterlin, R.A. 1974. Does economic growth improve the human lot, in *Nations and Households in Economic Growth: Essays in Honour of Moses Abramovitz*, edited by P.A. David and M.W. Reder. New York: Academic Press, 89-125.

Eckersley, R. 2009. Population measures of Subjective Wellbeing: How useful are they? *Social Indicators Research*, 94, 1-12.

Fleuret, S. and Atkinson, S. 2007. Wellbeing, health and geography: A critical review and research agenda. *New Zealand Geographer*, 63, 106-18.

Foley, R. 2010. *Healing Waters: Therapeutic Landscapes in Historic and Contemporary Ireland.* Farnham: Ashgate.

Frey, B.S. 2008. *Happiness: A Revolution in Economics.* Cambridge, MA: MIT Press.

Garvin, T. and Nykiforuk, C. 2011. *Ageing in Placelessness: Suburban Geographies of the Elderly.* Paper to the 14th International Medical Geography Symposium: Durham, UK, July 2011.

Gesler, W.M. 1992. Therapeutic landscapes: Medical issues in light of the new cultural geography. *Social Science & Medicine*, 34, 735-46.

Gesler, W.M. and Kearns, R.A. 2002. *Culture/Place/Health.* London: Routledge.

Gibson-Graham, J.K. 2006. *A Post-Capitalist Economics.* Minneapolis, MN: University of Minnesota Press.

Glasmeier, A. 2005. *An Atlas of Poverty in America: One Nation, Pulling Apart.* New York: Routledge.

Groenewegen, P.P., van den Berg, A.E., de Vries, S. and Veheij, R.A. 2006. Vitamin G: Effects of green space on health, well-being, and social safety. *BioMed Central Public Health*, 6(149).

Hall, E. 2007. Creating spaces of wellbeing for people with learning disabilities: A commentary. *New Zealand Geographer*, 63, 130-34.

Hartig, T., Evans, G.W., Jamner, L.E., Davis, D.S. and Gärling, T. 2003. Tracking restoration in natural and urban field settings. *Journal of Environmental Psychology*, 23, 109-23.

Harvey, D. 1972. Revolutionary and counter-revolutionary theory. *Antipode*, 4(2), 1-25.

Harvey, D. 1973. *Social Justice and the City.* London: Edward Arnold.

Helliwell, J.F. and Putnam, R.D. 2004. The social context of well-being. *Philosophical Transactions: Biological Sciences*, 359(1449), 1435-46.

Hillsdon, M., Panter, J., Foster, C. and Jones, A. 2006. The relationship between access and quality of urban green space with population physical activity. *Journal of the Royal Institute of Public Health*, 120, 1127-32.

Holmes, W., Stewart, P., Garrow, A., Anderson, I. and Thorpe, L. 2002. Researching Aboriginal Health: Experience from a study of urban young people's health and well-being. *Social Science and Medicine*, 54, 1267-79.

Huppert, F.A. 2009. A new approach to reducing disorder and improving well-being. *Perspectives on Psychological Science*, 4(1), 108-11.

Ingersoll-Dayton, B., Saengtienchai, C., Kespichayawattanna, J. and Aungsuroch, Y. 2004. Measuring psychological well-being: Insights from Thai elders. *Gerontologist*, 44, 596-604.

Izquierdo, C. 2005. When 'health' is not enough: Societal, individual and biomedical assessments of well-being among the Matsigenka of the Peruvian Amazon. *Social Science and Medicine*, 61, 767-83.

Kahneman, D. and Krueger, A.B. 2006. Developments in the measurement of subjective well-being. *Journal of Economic Perspectives*, 20(1), 3-24.

Kahneman, D., Diener, E. and Schwarz, N. (eds) 1999. *Well-being: The Foundations of Hedonic Psychology.* New York: Russell Sage Foundation.

Kane, S. and Wards, S. 1989. Social well-being in the New Zealand urban system: Inter- and intra-urban contrasts. *Proceedings, 15th New Zealand Geographical Society Conference.* Dunedin: University of Otago, 123-33.

Kaplan, S. 1995. The Restorative Benefits of Nature: Toward an Integrative Framework. *Journal of Environmental Psychology*, 15, 169-82.

Kearns, R. 1993. Place and health: Towards a reformed medical geography. *Professional Geographer*, 45, 139-47.

Kearns, R. and Andrews, G. 2010. Geographies of Well-Being, in *The Sage Handbook of Social Geographies*, edited by S.J. Smith, R. Pain, S.A Marston and J.P Jones III. London: Sage, 309-28.

Kearns, R. and Collins, D. 2010. Health Geography', in *A Companion to Health and Medical Geography*, edited by T. Brown, S. McLafferty and G. Moon. Oxford: Wiley-Blackwell, 15-32.

Kearns, R. and Gesler, W. 1998. *Putting Health into Place: Landscape, Identity and Well-being.* Syracuse, NY: Syracuse University Press.

Kearns, R. and Moon, G. 2002. From medical to health geography, novelty, place and theory after a decade of change. *Progress in Human Geography*, 26(5), 605-25.

Kearsley, G. 1982. Subjective social indicators and the quality of life in Dunedin. *New Zealand Geographer*, 38(1), 19-24.

Knox, P. 1974a. Level of living: A conceptual framework for monitoring regional variations in well-being. *Regional Studies*, 8, 11-19.

Knox, P. 1974b. Spatial variations in level of living in England and Wales in 1961. *Transactions of the Institute of British Geographers*, 62, 1-24.

Knox, P. 1975. *Social Well-Being: A Spatial Perspective.* Oxford: Oxford University Press.

Kulkarni, K.M. 1990. *Geographical Patterns of Social Well-being.* New Delhi: Concept Publishing.

Kuz, T.J. 1978. Quality of life, an objective and subjective variable analysis. *Regional Studies*, 12, 409-17.

Layard, R. 2005. *Happiness: Lessons from a New Science.* London: Penguin.

Luechinger, S. 2009. Valuing air quality using the life satisfaction approach. *Economic Journal*, 119, 482-515.

Maas, J., Verheij, R.A., Groenewegen, P.P., Vries, S. and Spreeuwenberg, P. 2006. Green space, urbanity, and health: How strong is the relation? *Journal of Epidemiology and Community Health*, 60, 587-92.

Maas, J., Verheij, R.A., Spreeuwenberg, P. and Groenewegen, P.P. 2008. Physical activity as a possible mechanism behind the relationship between green space and health: A multilevel analysis. *BioMed Central Public Health*, 8(206).

McKracken, K. 1977. Social well-being within the New Zealand urban system. *Proceedings, 9th New Zealand Geographical Society Conference.* Dunedin: University of Otago, 26-32.

McKracken, K. 1983. Dimensions of social well-being: Implications of alternative spatial frames. *Environment and Planning A*, 15, 579-92.

Milligan, C. and Conradson, D. (eds) 2006. *Landscapes of Voluntarism: New Spaces of Health, Welfare and Governance.* Bristol: Policy Press.

Mitchell, R. and Popham, F. 2007. Greenspace, urbanity and health: Relationships in England. *Journal of Epidemiology and Community Health*, 61, 681-3.

Mitchell, R., Astell-Burt, T. and Richardson, E.A. 2011. A comparison of green space indicators for epidemiological research. *Journal of Epidemiology and Community Health*.

Morrison, P. 2007. Subjective Wellbeing and the City. *Social Policy Journal of New Zealand*, 31, 74-103.

Morrison, P. 2010. Local expressions of subjective well-being: The New Zealand experience. *Regional Studies*, 45(8), 1039-58.

New Economics Foundation. 2008. *Five Ways to Wellbeing*. London: New Economics Foundation.

New Economics Foundation. 2010. *The Happy Planet Index 2.0*. London: New Economics Foundation.

New Economics Foundation. 2004. *A Well-being Manifesto for a Flourishing Society*. London: New Economics Foundation.

Pacione, M. 1980. Differential quality of life in a metropolitan village. *Transactions of the Institute of British Geographers*, 5, 185-206.

Pacione, M. 1982. The use of objective and subjective measures of quality of life in human geography. *Progress in Human Geography*, 6(4), 495-515.

Pacione, M. 1986. Quality of life in Glasgow: An applied geographical analysis. *Environment and Planning A*, 18(11), 1499-520.

Pacione, M. 2003. Urban environmental quality and human well-being: A social geographical perspective. *Landscape and Urban Planning*, 65, 19-30.

Panelli, R. and Tipa, G. 2007. Placing Well-Being: A Maori Case Study of Cultural and Environmental Specificity. *EcoHealth*, 4, 445-60.

Panelli, R. and Tipa, G. 2009. Beyond foodscapes: Considering geographies of Indigenous well-being. *Health and Place*, 15, 455-65.

Parr, H. and Philo, C. 2003. Rural mental health and social geographies of caring. *Social and Cultural Geography*, 4(4), 471-88.

Peet, R. 1975. Inequality and poverty: A Marxist-geographic theory. *Annals of the Association of American Geographers*, 65(4), 564-71.

Philo, C. 1995. *Off the Map: The Social Geography of Poverty in the UK*. London: Child Poverty Action Group.

Public Health Advisory Committee. 2010. *Healthy Places, Healthy Lives: Urban Environments and Well-Being*. Wellington, New Zealand: NZ Ministry of Health.

Rice, N. and Leyland, A.H. 1996. Multi-level modeals: Application to health data. *Journal of Health Services Research and Policy*, 1, 154-64.

Richardson, E., Pearce, J., Mitchell, R., Day, P. and Kingham, S. 2010. The association between green space and cause-specific mortality in urban New Zealand: An ecological analysis of green space utility. *BMC Public Health*, 10(1), 240.

Ryff, C.D. 1989. Happiness is everything, or is it? Explorations on the meaning of psychological well-being. *Journal of Personality and Social Psychology*, 57, 1069-81.

Scott, K. 2012. *Measuring Wellbeing: Towards Sustainability?* London: Earthscan.

Sen, A. 1992. Capability and wellbeing, in *The Quality of Life*, edited by A. Sen and M. Nussbaum. Oxford: Clarendon, 30-53.

Smith, D.M. 1994. *Geography and Social Justice*. Oxford: Blackwell.

Smith, D.M. 1973. *The Geography of Social Wellbeing in the United States: An Introduction to Territorial Social Indicators*. New York: McGraw-Hill.

Smyth, R., Mishra, V. and Qian, X. 2008. The environment and well-being in urban China. *Ecological Economics*, 68, 547-55.

Snidjers, T. and Bosker, R. 1999. *Multilevel Analysis: An Introduction to Basic and Advanced Multilevel Modelling*. London, Sage.

Sorensen, T. and Weinand, T. 1991. Regional well-being in Australia revisited. *Australian Geographical Studies*, 29(1), 42-70.

Stanca, L. 2010. The geography of economics and happiness: Spatial patterns in the effects of economic conditions on well-being. *Social Indicators Research*.

Stiglitz, J., Sen A. and Fitoussi, J., 2009. *Report by the Commission on the Measurement of Economic Performance and Social Progress* [Online]. Available at: www.stiglitz-sen-fitoussi.fr [accessed: 29 September 2009].

Stutzer, A. and Frey, B.S. 2008. Stress that doesn't pay: The commuting paradox. *Scandinavian Journal of Economics*, 110, 339-66.

Ulrich, R. 1984. View through a window may influence recovery from surgery. *Science*, 224, 420-21.

Veenhoven, R. 1996. Developments in satisfaction research. *Social Indicators Research*, 37, 1-46.

Vernon, M. 2005. *Well-Being*. London: Continuum.

Waldron, S. 2010. *Measuring Subjective Wellbeing in the UK*. London: UK Office for National Statistics.

Walmsley, D. and Weinand, H.C. 1997. Fiscal federalism and social well-being in Australia. *Australian Geographical Studies*, 35(3), 260-70.

Walton, D., Murray, S.J. and Thomas, J.A. 2008. Relationship between population density and the perceived quality of neighbourhood. *Social Indicators Research*, 89, 405-20.

Waterman, A.S. 1990. The relevance of Aristotle's conception of eudaimonia for the psychological study of happiness. *Theoretical and Philosophical Psychology*, 10, 39-44.

Watzlawick, P., Weakland, J. and Fisch, R. 1974. *Change: Principles of Problem Formation and Problem Resolution*. New York: Norton.

Welsch, H. 2006. Environment and happiness: Valuation of air pollution using life satisfaction data. *Ecological Economics*, 58, 801-13.

Wiles, J., Allena, R., Palmera, A.J., Haymana, K.J., Keeling, S. and Kerse, N. 2009. Older people and their social spaces: A study of well-being and attachment to place in Aotearoa New Zealand. *Social Science and Medicine*, 68, 664-71.

Williams, A. (ed.) 1999. *Therapeutic Landscapes: The Dynamic Between Place and Wellness*. New York: University Press of America.

Williams, A. (ed.) 2008. *Therapeutic Landscapes*. Aldershot: Ashgate.

Wilson, J.O. 1969. *Quality of Life in the United States: An Excursion into the New Frontier of Socio-Economic Indicators*. Kansas City, MO: Midwest Research Institute.

Wilton, R. and DeVerteuil, G. 2006. Spaces of sobriety/sites of power: Examining social model alcohol recovery programs as therapeutic landscapes. *Social Sciences & Medicine*, 63(3), 649-61.

Chapter 3

Understanding the Impact of Urban Green Space on Health and Wellbeing

Helen Beck

High quality, well managed and maintained green spaces are the backbone of sustainable and vibrant areas. They are a key ingredient of attractive, healthy and economically competitive places (CABE Space 2005d). They contribute positive social and environmental value and play a crucial role in promoting individual wellbeing (CABE 2010b, Greenspace Scotland 2008, Mitchell and Popham 2008). However, there is little accurate information available at a national level about England's green spaces and its relationship to our health and wellbeing.

In particular, it is information on the green spaces in towns and cities that is missing. Even basic data is absent. For instance, nobody knows how many green spaces there are, where they are, who owns them or what condition they are in. There have been very few large scale English studies that look at how health, green space and deprivation or ethnicity, are related and there is little significant research investigating income and race inequalities in urban green space provision and use. This has a serious and immediate impact on the ability to make robust arguments for the retention of green spaces in urban areas, and the value of devoting adequate funding to their upkeep and is a significant barrier to understanding the value of area quality in promoting people's health and wellbeing, and its role in combating social disadvantage.

To help fill this information gap, the Commission for Architecture and the Built Environment (CABE), the English government's national advisor on architecture, urban design and public space in urban areas, funded a programme of research between 2007 and 2010.[1] This aimed to develop the evidence base on the state of urban green space in England and its impact on people's health and wellbeing and paid specific attention to deprivation and ethnicity. The research was the first of its kind in England due to its scale and focus. Public green spaces were defined for the purpose of CABE's research as those spaces that are publicly owned, managed and maintained spaces that are, in theory, open and accessible to all. They include, for instance, parks, grass pitches, recreation grounds, woodlands and playgrounds. The World Health Organization's definition of health in terms of wellbeing was adopted: 'health is a state of complete physical, mental and social well-being

---

1   Research carried out by Heriot-Watt University and OPENspace Research Centre, Edinburgh College of Art.

and not merely the absence of disease or infirmity' (WHO 1946). In addition, the research drew on the work of McAllister (2005) which recognizes the need for objective and subjective measures of wellbeing, defining the main domains of wellbeing as: physical health, income and wealth, relationships, meaningful work and leisure, personal stability and lack of depression. The research drew on the domains known to have a relationship with green space – physical and mental health, relationships and meaningful leisure (McAllister 2005).

Deprived, ethnically diverse areas were the focus for a number of interrelated reasons. First, deprived communities experience the worst local environments. They are more likely to suffer from poorer air quality, higher levels of environmental crime and degraded public spaces (Environment Agency 2006). Historically, poor areas in towns and cities have been exposed to a larger share of environmental risks and dangers and in a changing climate they are also more likely to suffer disproportionately. Second, the relationship between low income and poor health follows a social gradient, where people living on a low income in deprived urban areas are more likely to experience worse health and be less physically active. The 2010 Marmot Review of Health Inequalities revealed that the gap in life expectancy between the rich and poor is greater in England than in three quarters of OECD countries (The Marmot Review 2010). Finally, some groups report worse health. Bangladeshi and Pakistani people for instance are more likely to report bad or very bad health compared to the general population (Joint Health Surveys Unit 2004). Most of the UK's black and minority ethnic communities live in England, in inner urban areas and in the most deprived wards with poverty rates double that found among white people (Platts 2006).

Public green spaces have a proven track record in reducing the impact of deprivation by delivering better health and wellbeing and creating a strong community. For example, research finds that, regardless of income, living in a greener environment reduces mortality (Mitchell and Popham 2008) and can help reduce the significant gap in life expectancy between rich and poor. Thus the research commissioned by CABE (CABE 2010a and 2010b) aimed to improve the available evidence around the impact of urban green space quality on health and wellbeing.

This chapter first looks at the current evidence about green spaces in England's urban areas. It discusses the available research that robustly proves the value that green spaces contribute to places, before moving on to set out CABE's programme of research. The chapter concludes with reflections on the position of green spaces in a changing political landscape and highlights key areas that would benefit from further research.

Green Space in England: A Meagre Evidence Base in its Infancy

Most of our towns and cities are endowed with a complex legacy of trees, parks, gardens, allotments, cemeteries, woodlands, green corridors, rivers and waterways.

This green infrastructure is multi-functional. It offers a working landscape, for instance soft landscape areas absorb heavy rainfall, and at the same time these spaces clean and cool the air and provide valuable space for exercise, play and socializing (CABE 2009). Understanding the multiplicity of values and benefit contributed by green spaces and its impact on people's health and wellbeing presents a number of interesting and slippery methodological issues. This chapter focuses on the issues that are specific to the collation and analysis of data about green spaces in urban England.

At present it is impossible to understand changes in the quantity and quality of urban green space in England over time. National data collection is inconsistent and patchy, and the information that is collected is often skewed towards specific, easy to measure targets. Data for some areas is more robust and comprehensive than it is for others. Some elements are measured over and over again, others are never measured. For instance, we know how clean and well maintained spaces are, but not how valuable, vibrant or functional they are (Heriot-Watt University and Oxford Brookes University 2007). Responsibility for data collection on landscape morphology and character, biodiversity and wildlife habitats, heritage and conservation is divided across at least three central government departments and there is little harmonization between existing data sources and no shared depository for this information. Information on green space is either not collected at a national level or is collated in different ways and formats locally.

On the whole, it is the sources of information on green spaces in rural areas that are the most developed whereas the urban evidence base is severely underdeveloped. For instance, the UK Land Cover Map does not at present include information on different types of urban green space, classifying urban areas as either suburban or urban. The Generalised Land Use Database includes two categories relating to green space: domestic gardens and green space, where the green space category covers anything green including farmland, parks and forest. Planning Policy Guidance 17 (PPG17) states that local authorities in England must carry out audits of existing open spaces, taking into account the use and access to these areas (Office for the Deputy Prime Minister 2002). However, the information has never been collated on a national scale and no standardized method of data collection is available, hindering comparability between areas. The reform of the planning system means that PPG17 is likely to be replaced by a new National Planning Policy Framework, coming into force in April 2012.

There are partial data sources on green spaces that can be drawn upon to indicate the extent of green space in urban areas. The Public Parks Assessment carried out in 2001 is the only national survey of parks and recreational open spaces in England (Urban Parks Forum 2001) and documents around 14,600 spaces (National Audit Office 2006). For the purpose of its research CABE assembled a single inventory of green space for English urban authorities by combining data from existing data sources and including records for more than 16,000 individual spaces (CABE 2010a).

However, some information simply does not exist at present. For instance, social landlords are responsible for the green and open spaces on the doorsteps of around four million households but are not required to collect data about their green and open spaces and generally do not chose to record their green space stock as separate information. This is a significant gap in information particularly as in some areas, such as London, the green spaces owned and managed by Housing Associations may be greater than the amount owned and managed by the local authority. Detailed information on staffing in the green space sector is also absent. The Census, Labour Force Survey and Annual Business Inquiry collects information on the number of employees in public administration who fall under the heading of 'skilled agricultural workers', a category which includes horticultural workers, gardeners and groundsmen/women as well as farmers and those working in agricultural or fishing trades. This information does not provide any information about staff employed by private contractors who work in public parks and open spaces. Neither does the available data tell us anything about the managerial or professional staff involved in parks and open space services (CABE 2010c). In sum, the evidence base remains meagre and underdeveloped. We especially lack credible sources of national evidence and indicators which consistently measure or monitor the value of green spaces on a larger scale and on a longitudinal basis.

A key barrier to data collection is that is not straightforward. Proving the economic, social and environmental value of urban green space is a complex and multifaceted area, complicated by the need to take into account long timescales as benefits can accrue over many years. Green spaces are by their nature multifunctional and as a result, analysis often falls between different academic areas. To date, cross disciplinary working into the many values presented by urban green space has been limited. Difficulty is compounded by the dearth of robust national data and the fact that green space value consists of elements that may never be easily measured due to the difficulty of controlling for interfering variables.

Robust Evidence Proving the Value of Green Space for Urban Areas

Despite the methodological quirks set out above, there are credible sources of evidence that robustly prove the value of green spaces for urban areas. CABE's publication *The Value of Public Space* provides a summary of how high quality parks and public spaces create economic, social and environmental value (CABE Space 2005d) and Greenspace Scotland's *Greenspace and Quality of Life* critically reviews research relating to the links between green space and a range of quality of life issues (Greenspace Scotland 2008). Research suggests that nearly everyone uses parks and green spaces in urban areas, and that this is highly valued. In England, 91 per cent of people report using parks to some extent and 42 per cent use these spaces at least once a week (CABE 2010a). Additionally the benefits of using green spaces are recognized and appreciated; 91 per cent of the public believe that public parks and open spaces improve quality of life and 74 per cent believe that parks and open spaces are important to health and mental and physical wellbeing (CABE 2004).

The value of green space for physical health is unquestionable. Good quality spaces will encourage people to make short journeys on foot or by bike (Sustrans undated). Regular physical activity contributes to the prevention and management of over 20 conditions including coronary heart disease, diabetes, certain types of cancer and obesity. For example, strokes cost the National Health Service £2.8 billion a year and physical activity reduces the risk of having a stroke by a third (National Heart Forum, Living Streets and CABE 2007). Access to green space also impacts positively on mental health. Responses to nature include feelings of pleasure and reduction in anxiety. Moderate activity is as successful at treating depression as medication (Natural England 2006). Healthy childhood development is also facilitated by green space. Children with access to safe green space are more likely to be physically active and less likely to be overweight (Greenspace Scotland, 2008) while outdoor play encourages healthy brain development and promotion of healthy wellbeing through adulthood (National Heart Forum, Living Streets and CABE 2007).

Furthermore, green spaces are critical for the adaptation to, and mitigation of, climate change impacts. They support biodiversity, absorb and manage surface water, mitigate the urban heat island effect and improve air quality; a role that will only get more important in future years (CABE Space 2008). Research by Gill et al. (2006) has shown that:

- increasing green space cover in urban areas by 10 per cent reduces surface run-off by almost five per cent.
- increasing tree cover in urban areas by 10 per cent reduces surface water run-off by almost six per cent.
- adding green roofs to all the buildings in town centres can reduce surface water run-off by almost 20 per cent.

Urban trees can also have a dramatic effect in reducing 'urban heat island' effects. For example, increasing tree cover by 25 per cent can reduce afternoon air temperatures by between six and 10°C (American Society of Landscape Architects 2006).

The quality of public space is an important factor in both attracting people to particular areas and retaining residents. In the British Household Panel datasets, respondents were asked to give reasons why their area was a good or bad place to live and 44 per cent of the reasons given related to quality of public space. Furthermore, the Survey of English Housing asked respondents to list the three main things that would improve their local area and issues relating to aspects of public space are cited as many times as factors relating to employment, health and housing (Heriot-Watt University and Oxford Brookes University 2007).

In recent years, greater attention has been paid to the financial contributions that high quality public spaces provide for the areas and cities within which they are located. City park economics is an emerging discipline. Research by the Trust for Public Land in America enumerates the economic value of the City of Philadelphia's park system for clean air, clean water, tourism, health, property values and community cohesion. The study calculates the financial benefits that parks in the city of Philadelphia contribute to their users as $1 billion (Trust for Public Land 2008). Similarly CABE Space research *Does Money Grow on Trees?* looks at how well-planned and managed parks, gardens and squares can have a positive impact on the value of nearby properties and can attract inward investment and people to an area. The increase in value ranged from between 0 to 34 per cent, with a typical increase of about 5 per cent (CABE Space 2005b).

Research commissioned by Natural Economy Northwest, a joint programme of the Northwest Regional Development Agency and Natural England, brings together a wide range of evidence on the multiple benefits of green infrastructure, focusing in particular on its role in creating economic prosperity and stability for the region. The research calculates that the Northwest's environment generates an estimated £2.6 billion in gross value added and supports 109,000 jobs in environment and related fields (Natural Economy North West 2008). Conversely, neglected, poorly managed and maintained spaces create negative value – contributing to the onset of vandalism, anti-social behaviour, graffiti and rubbish (CABE Space 2005a). Neglected spaces create the impression that areas are not cared about and contribute to circles of decline whereas successful, quality spaces can be the making of a place and attract people to live, work, visit and invest in a particular area (CABE Space 2005c).

Understanding More About the State of England's Urban Green Spaces

CABE's research set out to robustly expand the available evidence base by gauging the state of England's urban green spaces and exploring why it matters for people's health and wellbeing. The research was carried out in two parts and published as two separate studies: *Urban Green Nation: Building the Evidence Base* and *Community Green: Using Local Spaces to Tackle Inequality and Improve Health.*

Urban Green Nation is the first review of the urban evidence base and took the form of an extensive literature review and analysis of relevant data. Existing national level data relating to green space in urban areas in England was compiled and analysed by Heriot Watt University on behalf of CABE. Over 70 major data sources were investigated and an inventory of over 16,000 individual green spaces assembled. The study did not consider privately owned green spaces such as communal or private gardens or the grounds of institutions such as universities and art galleries. Instead, it concentrated on publicly owned, managed and maintained spaces that are, in theory, open and accessible to all. The data was analyzed by six themes, selected to represent as far as possible using existing data a multi-faceted view of green space:

1. Quantity: by type of green space, including both absolute and relative amounts, available in urban areas.
2. Quality: including subjective assessments such as resident satisfaction and objective measures such as biodiversity.
3. Use: how people use green space.
4. Proximity: the physical location of green space in relation to where people live, and how far people have to travel to access different types of green space.
5. Management and maintenance: including information about spending, staffing and how well a space is looked after.
6. Value: capturing how important green space is to people.

The research particularly focused on connections between different aspects of green space and the local environment, taking account of wider socio-demographic factors, location, housing density and other issues. The analysis found that people are using their urban parks and open spaces more, and they value these spaces (CABE 2010a). Almost nine out of 10 people use their local green spaces and parks and open spaces are the most frequently used service of all the services tracked by The Place Survey.[2] This compares with 32 per cent for concert hall visits and 26 per cent for galleries. In fact, research reports 1.8 billion visits to parks in England every year (Heritage Lottery Fund, 2009). This appreciation for local spaces is increasing: in 2007 91 per cent of people thought it very or fairly important to have green spaces near to where they live but by 2009 this had risen to 95 per cent (DEFRA 2007, 2009).

Region-by-region data analysis revealed some interesting general trends in quality and quantity. The quantity of urban green space varies considerably between the government regions and types of urban location. The South West, South East and East Midlands tend to have higher levels of green space provision, compared with London, the North West and the West Midlands. Suburban and town fringe areas tend to have more public open space and green space than city centres, although city centres tend to have more recreation facilities and play areas. Furthermore, quality indicators are generally more favourable in the southern regions compared with the northern regions, and generally better in suburban than in urban/city areas, except for central London.

However these differences in regional provision are less significant than the differences in provision according to the level of deprivation or affluence of a particular place. The quality of local green spaces differed dramatically according

---

2   The Place Survey was published in 2009 and was a Department of Communities and Local Government survey of resident's satisfaction with local authority provided services and neighbourhood satisfaction.

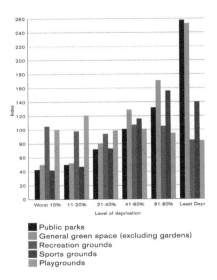

Public parks
General green space (excluding gardens)
Recreation grounds
Sports grounds
Playgrounds

Figure 3.1   Quantity and type of green space and area deprivation

to people's socio-economic and cultural background and people in deprived areas, wherever they live, receive a far worse provision of parks and green spaces than their affluent neighbours. This is significant as these groups often do not have gardens and access to good quality public green space matters more.

The most affluent 20 per cent of wards in urban England have five times the amount of parks and general green space (excluding private gardens) than the most deprived 10 per cent of wards. Therefore if you live in an affluent suburb, you are also likely to have an above-average quantity of good parks nearby that are more likely to have a dedicated budget and staff allocated for their management and maintenance. On the other hand, if you live in a deprived inner-city ward, with high-density housing, you might have many small, poor-quality green spaces but you are unlikely to have access to large green spaces, or good-quality green space. Comparing deprived and affluent areas, residents' general satisfaction with their neighbourhood falls from around 80 per cent in affluent places to around 50 per cent in the most deprived places.

Analysis also showed that people from minority ethnic groups have less local green space and that which exists is of poorer quality. Although where you live is intimately related to income, the research found a difference by ethnicity that was over and above what would be expected for level of income alone. Areas that have almost no black and minority ethnic residents (fewer than two per cent of ward population) have six times as many parks as wards where more than 40 per cent of the population are people from black and minority ethnic groups. They have 11 times more public green space if all types (excluding gardens) are looked at. In addition, the data shows that the quality of green space available to black and minority ethnic people is also worse.

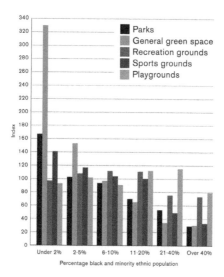

Figure 3.2   Quantity and type of space by black and minority ethnic population

An area's public green spaces are a public service that nearly everyone uses. Outside of their direct use, everybody benefits from the wider values contributed by green spaces – be it from cleaning of the air or the reduction of the urban heat island effect. The discrepancy in provision and quality set out above matters. Some people must manage a greater number of burdens yet have fewer environmental assets or resources to draw upon. Planning for the future must take this into account to ensure some areas are not more likely to be hazardous to health and wellbeing than others.

Using local spaces to tackle inequality and improve health

Over the last decade, promoting equal access and participation with respect to green space and the natural environment has emerged as a key environmental justice concern in the UK. A growing number of small scale research studies in the UK have examined the benefits of specific high quality public spaces and the association of these benefits with specific dimensions of quality of life. However, research to date has largely focused on the quantity of green space in relation to deprivation. There remains a big gap in the literature in relation to the quality and access to green space, especially in relation to deprivation and ethnicity.

Taking the results reported by *Urban Green Nation* as its starting point, the second part of CABE's research developed this evidence base further to investigate the inter-relationship between urban green space, inequality, ethnicity and health and wellbeing (CABE, 2010b). Carried out by OPENspace Research Centre,

Edinburgh College of Art, in collaboration with Heriot-Watt University, it is the largest study of its kind in England and asked:

1. How significant is the quality of urban green space is to the health and wellbeing of different socio-economic and ethnic communities living in six deprived urban areas of England?
2. What is the impact of varying quality of urban green space on health and wellbeing in these areas?

A literature review on urban green space, deprivation and ethnicity and its contribution to health and wellbeing was undertaken. This was supplemented by a review of 50 projects engaging people in the design, ownership and management of local green space in urban areas. Overall, the literature review found a lack of quantitative research using large sample sizes. Instead, most of the research reviewed took a case study approach – using qualitative methods in small samples, focusing on adults rather than on children. Six case study areas were chosen in London, the West Midlands and Greater Manchester, because of their high level of deprivation, high percentage of black and minority ethnic populations and because they contained green spaces of different levels of quality.[3] Information on the quantity and quality of green space in the areas was drawn from the *Urban Green Nation* study.

The study used the World Health Organization's definition of wellbeing: "health is a state of complete physical, mental and social wellbeing and not merely the absence of disease or infirmity" (WHO 1946). In addition, the research drew on the work of McAllister which recognizes the need for both objective and subjective measures of wellbeing. *Community Green* focused on physical and mental health, relationships and meaningful leisure, as these are the areas known to have a relationship with access to and use of green space.

The findings from the literature review informed the development of a household survey which asked how important interviewees thought access to green space was in relation to other factors in making an area 'a good place to live'. In order to do this, the survey drew on a questionnaire developed as part of a European Union-funded project entitled PLUREL (Peri-urban Land Use Relationships: Strategies and Sustainability Assessment Tools for Urban – Rural Linkages).[4] PLUREL examined how important access to green space is in relation to seven other environmental attributes:

- Air quality.
- Suitability of housing.
- Area safety and security.

---

3   Indices of multiple deprivation data. Areas chosen from the top 20 per cent of deprived neighbourhoods.

4   PLUREL is a large research project (2007-2010) funded within the 6th Research Framework Programme of the European Union. Available at: www.plurel.net.

- Noise pollution.
- Shopping facilities.
- Public transport.
- Waste disposal.

These attributes are considered to be a manageable set of physical environmental factors that are likely to be pertinent to people's wellbeing and are most relevant in making a neighbourhood a good place to live. Therefore, using PLUREL as its basis, the household survey explored the relative importance of urban green space in relation to other environmental attributes using a computerized simulated exercise whereby participants were asked to select what would make a good place to live.

Secondly, interviewees were asked about their health, their use of green space, the quality of their local green spaces and how improvements to their local spaces would affect their use and levels of physical activity. This aimed to gain an insight into how improvements in the quality of local green space impact on people's use of, and levels of activity within, these spaces, and consequently their health and wellbeing. There were few precedents in previous research for exploring how perceived wellbeing changes in relation to green space. A total of 523 interviews with white British (22 per cent of interviewees), Pakistani (22 per cent), Bangladeshi (17 per cent), black African and African-Caribbean (12 per cent) and Indian people (11 per cent) were carried out.[5] Overall, people were very willing to talk about their local green spaces, especially those households with children, with a 68 per cent response rate.

The study found that everyone views green space as a key and basic service, alongside housing, education and policing. Overall, area safety and security were considered most important, within the context of the attributes evaluated, contributing 16 per cent towards making an area 'a good place to live'. This result varied in importance according to people's ethnicity, with Bangladeshi, black African and African-Caribbean people rating safety as the second most important attribute after the design and construction of housing. Access to green space was ranked sixth in importance and contributed around 10 per cent towards what made an area a good place to live.

The green space that mattered most to people in our study was the local park, which received a resounding vote of confidence despite varying levels of quality and use. It accounted for 90 per cent of the green spaces all people used. The majority of interviewees (78 per cent) visited their nearest green space on foot, indicating that it is the local neighbourhood park that is of most significance as a resource that people choose to use. However, in locations with a higher-quality park, for instance with a Green Flag award, people did travel further.

---

5 Ethnicity was self-reported. 16 per cent of people were from other black and minority ethnic groups that included dual heritage people, Chinese and Turkish people. African-Caribbean and black African interviewees were combined into one group for analysis due to small study numbers.

A person's ethnicity was the strongest indicator of their green space use in the survey data. Some groups visited green spaces more than others, and for different purposes, with physical activity and social patterns of use generally more popular among black and minority ethnic interviewees. Patterns of use differed most between Bangladeshi interviewees, who were more likely to visit to get fresh air, meet friends and for physical activity, and all other interviewees, particularly white British people who were more likely to visit green space for relaxation and to enjoy the peace and quiet of the space.

Barriers to using local green spaces reported by interviewees included feelings of insecurity due to the fear of personal attack or racism, exclusion due to the domination of a space by a particular group and the presence of dogs (dog-fouling or fear of dogs). Having nowhere to sit was also cited as a barrier to use. The biggest single reported barrier was perceptions of safety. This differed by ethnicity as only half of Bangladeshi people, for example, reported feeling safe using their local green space, compared with three quarters of white people interviewed. Other statistical differences were found by ethnicity; for instance, white British and Indian interviewees reported litter and dog fouling as major barriers to use.

People mentioned the loss of well-used and valued facilities such as football and cricket pitches as reasons why they used space less. This was especially true of young people. Aspirations for good-quality green space were common to all interviewees. Everybody wanted more facilities such as cafés, toilets, play and sports provision, and improved safety and more community events. 46 per cent of people said they would use their local green space more often if it had better facilities. Overall, if their local green space were made more pleasant and they began to use it more, 60 per cent of interviewees thought it would improve their overall physical health, 48 per cent thought it could improve their mental health, and 46 per cent thought it would make them feel better about their relationships with family and friends. Indian interviewees reported the highest perceived benefits of better local green space.

While physical activity did not feature highly in people's current use patterns, in terms of future use (based upon an improved-quality green space), it featured much more highly. 52 per cent of those asked said they would do more physical exercise if green spaces were improved. Indian, Bangladeshi and Pakistani people were more likely than other ethnic groups to visit urban green space for exercise. This suggests that improved green space use by these groups would also be more active use, and could make an important contribution to better health in black and minority ethnic groups.

Yet significant local green space resources remain unexploited. Public parks are far from being the only green spaces in towns and cities, yet they accounted for 90 per cent of the green spaces all people used. Less than one per cent of those living in social housing (21 per cent of our interviewees) reported using the green spaces in the housing estate they live in. This may be due to concerns about safety, lack of access or poor quality. Thus the green spaces around their homes are a critical, latent resource for health and wellbeing that are not being used to their full potential.

Conclusion and Discussion

It was far sighted Victorian civic leaders and philanthropists who recognized the health, social and environmental values of public parks in providing a place for people to meet, exercise and in forming the green lungs of our towns and cities. Our green and open public spaces are probably the only public service able to provide so many multiple, concurrent public benefits to the specific areas within which they are located. For example, where people perceive green space quality to be good, they are more satisfied with their neighbourhood and are more likely to report better health (CABE 2010b). Today, with regard to urban green space quality, it is still deprived areas and those with high black and minority ethnic populations that systematically fare worse in nearly all respects – particular in terms of quantity, quality and level of use.

It has never been more important to prove and robustly argue the worth of the green spaces in towns and cities. After 10 years of attention the historic decline of public urban green spaces was arrested but even during this period of relative prosperity, not everywhere benefitted equally. The landscape of public management for England's parks and green spaces is fundamentally changing again. While local authorities are being given greater freedom to define their contribution to local communities, it is accompanied by significant reductions in the resources available. The changes asked of public services go further than identifying efficiency savings. In most cases, substantial restructuring and radically different approaches to service delivery will accompany challenging decisions about what the public sector will provide. The provision of green space is not a statutory requirement.

In this context it is critical to make the best use of existing evidence to increase understanding of the benefits of green space to communities and its integral role in both area success and decline. Inevitably, things that are easily definable and understood tend to get the most attention. Those things that cannot be easily measured can be more easily overlooked and undervalued. Severe cuts to the management or maintenance of green spaces could be a false economy, producing costs in other areas, for instance policing the effects of anti-social behaviour in a derelict space.

There are opportunities. Interventions on places can reach many people simultaneously. For example, improving the provision of green space will benefit everybody and is a good way to target inequality. An obvious opportunity highlighted by *Community Green* is improving the quality of open spaces on social housing estates to tackle inequality, improve the wellbeing of residents and facilitate stronger and more cohesive places. The desire to improve local spaces is a powerful cohering force. CABE's research demonstrates the need to avoid making assumptions about what people, and different groups, want from their spaces. 'One size fits all' green space does not work: only flexible spaces will meet the needs of a diverse community. The unusually high response to the *Community Green* household survey demonstrated the interest people have in their local green space.

As the ethnic and age profile of the UK changes, green space managers need to understand the attitudes, needs and different reasons for green space use among local groups. They must work harder to involve their community in the management, planning and delivery of spaces. Active marketing of sites; events and activities such as community fun days; guided walks; space for allotments; and considering alternative uses of specific areas will all bolster usage and result in a healthier and more satisfied community. There are at least 4,000 'friends of' parks groups, and innumerable other voluntary community organizations established because people want a say over their local spaces. They are among the most passionate organizations, contributing around *£35 million a year in-kind economic value and leading significant change* (GreenSpace 2003). This is a powerful source of knowledge, energy, enthusiasm and front line support for those responsible for the management and maintenance for spaces.

There is great scope for further research on the benefits of urban green space. Proving the environmental benefits of urban green space is an emergent science and most focus to date has been on rural areas. The value of green infrastructure, for instance the networks of parks, gardens, allotments, trees, green roofs, cemeteries, woodlands, grasslands, moors and wetland areas, for towns and cities requires further analysis. This should consider issues at a wider scale than has been employed to date. For instance the environmental value of green infrastructure for the management of flood water is wider than the quantity of water that is stored, but also about the operation of green networks across different spatial scales.

The relationship between access and use of green space for positive health outcomes is explanatory, not causal. More work is needed on children's use and access to green space. Children have less contact with nature now than at any time in the past and it is estimated that by 2020 half of all children could be obese (The Strategy Unit 2008). There is a lack of in-depth investigation into deprivation, ethnicity and the quality and types of access to urban green space. Evidence of income and race inequalities in access to urban green space in the UK is limited to a handful of studies and most of the research on ethnicity and landscape has focused on rural contexts. However, access to nature mostly occurs in the local, urban neighbourhood context as black and minority ethnic populations mainly live in inner city and urban areas. There is also a lack of quantitative research targeted at larger samples of black and minority ethnic groups in relation to health and physical behaviour and attitudes to green space. Furthermore, research exploring how urban green space facilitates social integration and community cohesion is limited.

This is an exciting and valuable area with much to gain from further drilling down of associations and piecing together analysis of existing evidence. Clearer, more robust information is certainly required to justify attention. At the same time, it could be argued that in a sensible world this is a truth that does not need to be 'proved'. We instinctively know that access to good quality public green and open spaces provides benefits. Effort needs to be dedicated to sensibly growing the relevant pool of evidence but it must be acknowledged that it is not possible to 'prove' everything. Sometimes we must make the leap of faith without the absolute proof.

References

American Society of Landscape Architects. 2006. *Green Infrastructure* [Online]. Available at: http://www.asla.org/ContentDetail.aspx?id=24076 [accessed: 28 October 2010].

CABE Space. 2005a. *Decent Parks? Decent Behaviour? The Link between the Quality of Parks and User Behaviour.* London: CABE.

CABE Space. 2005b. *Does Money Grow on Trees?* London: CABE.

CABE Space. 2005c. *Start with the Park: Creating Sustainable Urban Green Spaces in Areas of Housing Growth and Renewal.* London: CABE.

CABE Space. 2005d. *The Value of Public Space: How High Quality Parks and Public Spaces Create Economic, Social and Environmental Value.* London: CABE.

CABE Space. 2008. *The Role of Public Space in Adapting to Climate Change.* London: CABE.

CABE Space. 2009. *Making the Invisible Visible: The Real Value of Park Assets.* London: CABE.

CABE. 2009. *Grey to Green: How We Shift Funding and Skills to Green our Cities.* London: CABE.

CABE. 2004. *Public Attitudes to Architecture and Public Space: Transforming Neighbourhoods* (unpublished).

CABE. 2010a. *Urban Green Nation: Building the Evidence Base.* London: CABE.

CABE. 2010b. *Community Green: Using Local Spaces to Tackle Inequality and Improve Health.* London: CABE.

CABE. 2010c. *Green Space Skills 2009: National Employer Survey Findings.* London: CABE.

DEFRA. 2007. *Survey of Public Attitudes and Behaviours Towards the Environment.* London: DEFRA.

DEFRA. 2009. *Public Attitudes and Behaviours Towards the Environment – Tracker Survey.* London: DEFRA.

Environment Agency. 2006. *Creating a Better Place: Environment Agency Corporate Strategy 2006-2011.* Bristol: Environment Agency.

Gill, S., Handley, J., Ennos, R. and Pauleit, S. 2006. Adapting Cities for Climate Change: The role of the green infrastructure. *Built Environment*, 3(1), 115-33.

Greenspace Scotland. 2008. *Greenspace and Quality of Life: A Critical Literature Review.* Stirling: Greenspace Scotland.

GreenSpace. 2003. *Community Networking Project.* Reading: GreenSpace.

Heriot-Watt University and Oxford Brookes University. 2007. *Understanding the Links Between the Quality of Public Space and the Quality of Life: A Scoping Study.* London: CABE.

Heritage Lottery Fund. 2009. *HLF Funding for Public Parks 1 April 1994 – 31 March 2009.* Heritage Lottery Fund Policy and Strategic Development Department Internal Data Briefing, October 2009.

Joint Health Surveys Unit. 2004. *Health Survey for England: The Health of Minority Ethnic Groups.* Leeds: The Information Centre.

McAllister, F. 2005. *Wellbeing Concepts and Challenges: Discussion Paper* [Online]. Available at: http://www.sd-research.org.uk/wp-content/uploads/sdrnwellbeingpaper-final_000.pdf [accessed: 28 October 2010].

Mitchell, R. and Popham, F. 2008. Effect of exposure to natural environment on health inequalities: An observational population study. *The Lancet*, 372(9650), 1655-60.

National Audit Office. 2006. *Enhancing Urban Green Space.* London: ODPM.

National Heart Forum, Living Streets and CABE. 2007. *Building Health: Creating and Enhancing Places for Healthy Active Lives.* London: National Heart Forum.

Natural Economy North West. 2008. *The Economic Value of Green Infrastructure* [Online]. Available at: www.naturaleconomynorthwest.co.uk/green+infrastructure.php [accessed: 28 October 2010].

Natural England. 2006. *Physical Activity and the Natural Environment.* Sheffield: Natural England.

Office for the Deputy Prime Minister. 2002. *Planning Policy Guidance 17: Planning for Open Space, Sport and Recreation.* London: ODPM.

Platts, L. 2006. *Poverty and Ethnicity in the UK.* York: Joseph Rowntree Foundation.

Sustans. Undated. *The Evidence!* [Online]. Available at: www.sustrans.org.uk/what-we-do/active-travel/active-travel-information-resources/the-evidence [accessed: 28 October 2010].

The Marmot Review. 2010. *Fair Society, Healthy Lives: Strategic Review of Health Inequalities in England Post-2010.* London: The Marmot Review.

The Strategy Unit. 2008. *Food Matters: A Strategy for the 21st Century.* London: Cabinet Office.

Trust for Public Land. 2008. *How Much Value Does the City of Philadelphia Receive from its Parks and Recreation System?* Philadelphia, PA: Trust for Public Land.

Urban Parks Forum. 2001. *Public Parks Assessment: A Survey of Local Authority Owned Parks Focusing on Parks of Historic Interest.* Caversham: Urban Parks Forum.

WHO. 1946. *Preamble to the Constitution of the World Health Organization* as adopted by the International Health Conference, New York, 19-22 June, 1946; signed on 22 July 1946 by the representatives of 61 States (Official Records of the World Health Organization, no. 2, 100) and entered into force on 7 April 1948.

Chapter 4

The Significance of Material and Social Contexts for Health and Wellbeing in Rural England

Mylène Riva and Sarah Curtis

In this chapter, we mainly discuss 'good health' (and absence of ill-health) which we consider here as possible dimension, or close correlate, of 'wellbeing'. We examine evidence concerning the ways that 'good health' experienced by individuals relates to the social and material context of their geographic living environment, with a focus on health in rural areas of England. To do so, we rely on individual health data from the Health Survey for England (HSE) linked to information pertaining to material deprivation and social cohesion in the small geographic areas of residence of survey respondents. Before delving into this quantitative account, we first present an overview of the literature on the associations between material and social conditions, rurality of places and experiences of good health and wellbeing.

Wellbeing, Health and Place

Wellbeing and health are interconnected: wellbeing contributes to good health and good health contributes to, and is a component of, wellbeing (Fernandez-Ballesteros et al. 2001, Fleuret and Atkinson 2007). The definition of health adopted in 1946 by the World Health Organization (WHO) draws attention to the multidimensional and positive nature of health, recognizing that health is 'a state of complete physical, mental and social wellbeing and not merely the absence of disease or infirmity' (WHO 1946). Two hallmarks of this definition are that it is holistic, recognizing the linkages between physical states and both emotional and social experiences, and that it strives towards positive states of wellbeing, rather than the absence of disease. This definition of health raises interesting issues about the links between health and other aspects of wellbeing, particularly since some interpretations suggest that the means to 'health' may be a fundamental human right (hence the generally accepted responsibility of governments to address the 'wider determinants of health' discussed below) (WHO 1978, Panter-Brick and Fuentes 2010) whereas social responsibility for individual's 'wellbeing' may be more contentious. Although it is outside the scope of this chapter to undertake a detailed discussion of the relationships between health and wellbeing, our discussion here should be considered in relation to other chapters in this book which explore meanings of wellbeing. We take the position in this chapter that the wider determinants of health are also likely to contribute to wellbeing, and we are particularly concerned here with whether geographical social and material conditions, varying locally across rural as well as urban areas, are important for the variations in self-reported health and psychological distress.

Moving away from the traditional biomedical model of health, it is now understood that health is only partly explained by 'conventional' behavioural factors (e.g. smoking) and other individual risk factors (e.g. low socioeconomic status) (Davey-Smith et al. 1990, Lynch et al. 1997). It is increasingly appreciated that health and wellbeing also depend significantly on 'wider determinants' (WHO Commission on the Social Determinants of Health 2008) and that the socioeconomic and material conditions of residential places also play an important role. Place constitutes, as well as contains social relations that are relevant for health and wellbeing (Kearns 1993, Macintyre et al. 1993, Curtis and Jones 1998, Macintyre et al. 2002, Curtis 2004). Places may vary in the extent that they provide opportunities, resources and/or constraints for people to live healthy lives. Thus place has come to be viewed as a structural determinant of health (Frohlich and Potvin 2008). In recent years, an interdisciplinary research agenda on 'place and health' has emerged at the boundaries of epidemiology and social sciences, focussed on the social and material characteristics of geographic life environments as potential risk factors for health (Jones and Moon 1993, Kearns 1993, Kearns and Joseph 1993, Macintyre et al. 1993, Berkman and Kawachi 2000). This research often tries to establish whether places have significance for health variation over and above the socioeconomic status and lifestyles of individual people living in these places, and, more recently, has concentrated especially on the interactions between individual and place characteristics, in order to be more sensitive to the diversity of individual responses to varying environments.

A growing body of evidence is accumulating (Riva et al. 2007) reporting significant effects of the social and material environmental conditions of residential settings on a variety of health outcomes including cardiovascular diseases and risk factors (Chaix 2009), mortality and self-rated health (Riva et al. 2007), mental health (Kim 2008, Mair et al. 2008, Curtis 2010), over and above the characteristics of local people that might be associated with health. This evidence has critical implications for the surveillance, protection, prevention and promotion functions of public health. Indeed, increasing knowledge about the etiological significance of places for health is useful for strategies designed to reach large groups of people (even entire populations) by changing environmental conditions to make them more conducive to health and wellbeing (Patychuk 2007). Academic discourse on these issues is paralleled by a wider public debate (e.g. Stratton 2010) on the need to measure the 'wellbeing' of the population as an alternative to economic indicators of human progress, and on the value for wellbeing of factors at the community level, such as neighbourliness and community empowerment (Young Foundation 2008).

The social, demographic, economic and environmental factors influencing health status are geographically distributed across space. Furthermore, there is also evidence of geographic disparity in health between metropolitan regions and rural areas in many countries. Whereas we observe a relative health *disadvantage* for rural populations in larger, more sparsely populated countries such as Canada, the United States and Australia (Eberhardt and Pamuk 2004, Pampalon et al. 2006, Australian Institute of Health and Welfare 2007), in Western European countries, the reverse is observed: on average, rural populations enjoy higher levels of health and wellbeing. For example, in the UK, it has been reported that rural dwellers enjoy longer life expectancy (Kyte and Wells 2010), lower rates of all-cause mortality (Levin 2003, Gartner et al. 2008) and better self-reported measures of physical and mental health (Kelleher et al. 2003, Weich et al. 2006, Riva et al. 2009). This said, for certain health outcomes, populations of rural areas have relatively poor health indicators compared to urban areas. In the UK and Ireland, this is the case, for example, for mortality from suicide (Levin and Leyland 2005, Middleton et al. 2006) and accidents (Boland et al. 2005, Gartner et al. 2008). In addition to urban-rural health disparities, some studies have reported geographic disparities in health between different types of rural areas (Gartner et al. 2008, Riva et al. 2009) and social inequalities in health within rural areas (Kyte and Wells 2010, Riva et al. 2011).

To date, most studies have focused on ill-health in deprived urban settings. One reason for this is the greater concentration of disadvantaged populations living in large urban centres suffering from relatively poor health and life expectancy, with attention often focused on the disparities between deprived and affluent areas within cities (Hartley 2004). Yet rural areas also show health inequalities (discussed below) and the health and wellbeing of rural populations may be vulnerable to social and economic changes taking place nationally and internationally, as well as locally. Improved knowledge of the conditions that foster good health in rural places is therefore crucial to efforts to improve and maintain wellbeing in these locations.

Research on place and health shows that differences in health and wellbeing between urban and rural areas might be interpreted partly in terms of the *composition* of the area, i.e. by the different 'types' of people living in urban vs. rural areas, differentiated by social and demographic factors that are important for health. The socio-demographic profile of rural England is rapidly changing. Rural areas account for about 19 per cent of the English population and are growing faster than urban areas, largely due to internal migration within the country (Commission for Rural Communities 2010). Wealthier people are migrating to rural areas, leading to a gentrification of the countryside, where less affluent households are increasingly being marginalized in the rural housing market. However, the relative disadvantage of poor individuals is not always apparent in rural areas, since they are not concentrated in such large numbers as in deprived urban areas, and information on rural inequalities is averaged over heterogeneous local communities (Haynes and Gale 2000).

Over and above individuals' socioeconomic position, disparities in health and wellbeing might also be explained by the *context* of urban and rural areas (geographic inequalities in health), i.e. material and social conditions of local communities, including deprivation levels and social cohesion. Collective resources, such as higher quality services and amenities, wealth, employment opportunities and social support, provide opportunities for people to live healthy lives and may be especially important for poorer people as they may be more reliant on collective resources and services. Community cohesion, seen as the social organization of a community, might confer positive benefits for health and wellbeing by reducing stress levels and providing practical help, through mechanisms such as perceived trust and reciprocity among community members. Area-level deprivation and social cohesion have also been reported as important determinants of various health outcomes in a range of settings, including self-rated health (Riva et al. 2007) and mental health (Kim 2008, Mair et al. 2008). A 'collective resources model' thus suggests that people living in areas characterized by more and better social and material collective resources enjoy better health and wellbeing than other groups living elsewhere (Stafford and Marmot 2003). In England, rural areas are, on average, less deprived and may, in some respects, be more socially cohesive than urban areas (Commission for Rural Communities 2008). These more favourable collective resources might explain the health advantage observed in rural areas.

Figure 4.1 presents the distribution of deprivation using the 2004 Index of Multiple Deprivation (IMD) (Noble et al. 2004) and the distribution of social cohesion using the Index of Social Fragmentation based on data from the 2001 Census (Congdon 1996) for small areas defined using the boundaries of Middle Super Output Areas (MSOAs; this geography is further explained below). These MSOAs are characterized as urban areas located either in the Greater London area or in 'other cities', and as rural areas defined as small towns or villages. Lower deciles correspond to lower levels of deprivation and social fragmentation (left side of the histograms), and higher deciles to higher levels of deprivation and social fragmentation (right side of the histograms). As can be seen, a greater proportion of urban MSOAs are categorized as falling in deciles of higher deprivation and social fragmentation, especially for those located in the Greater London area, with fewer MSOAs in deciles related to more favourable material and social contextual conditions of areas. Conversely, most rural MSOAs fall within more favourable deciles of deprivation and social fragmentation, with very few rural MSOAs being categorized as having very poor material and social contextual conditions in comparison to the national average.

Even in places with generally better social and material collective resources, individuals who are relatively disadvantaged may see that their own socioeconomic position and living conditions compare unfavourably with those of their neighbours (Stafford and Marmot 2003). The psychological effects of comparing oneself to wealthier neighbours may be damaging for health and wellbeing (Stafford and Marmot 2003, Wilkinson and Pickett 2007). In rural England, the socioeconomic

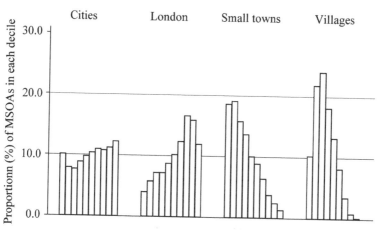

**Distribution of deciles of the Index of Multiple Deprivation 2004
(from least deprived to most deprived deciles)**

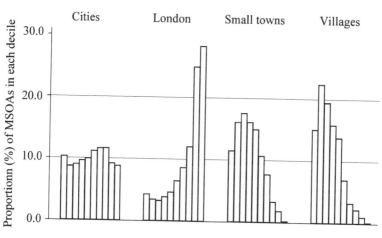

**Distribution of deciles of the Index of Social Fragmentation
(from least deprived to most deprived deciles)**

**Figure 4.1   The urban and rural deprivation profiles of small areas
(Middle Super Output Areas) in England**

Sources: Noble et al. (2004). Indices of deprivation 2004 and Congdon (1996), data from the 2001 census. Graphs produced by authors.

Note: For each group of MSOAs, areas are displayed in order of deprivation/fragmentation from least deprived/fragmented on the left to most deprived/fragmented on the right.

profile of rural populations is heterogeneous, often characterized by wealthier individuals living in close proximity to people struggling to make ends meet (Haynes and Gale 2000). Thus area indicators of rural health advantage may mask more local or individual inequalities in health and wellbeing. Individuals' responses to local conditions may also vary, depending on their own vulnerabilities or resilience.

In the following discussion, we empirically explore how place relates to wellbeing in England, with a focus on urban-rural and intra-rural disparities in indicators of good health that probably have a close association with wellbeing, though they focus more precisely on positive self-reported health and on absence of psychological distress. We aim to address the following research questions:

1. How do measures of positive self-reported health and on absence of psychological distress compare between urban and rural areas?
2. Are urban-rural differences in these health measures 'explained' by the characteristics of individuals living in rural areas and/or by the social and material conditions at the community level?
3. Do these health indicators vary more locally among types of rural settlements and if so is this associated with socioeconomic attributes of individuals and areas?
4. How do health inequalities at local level within rural areas compare with urban health inequalities?

In the discussion below we draw upon our previous work on health inequalities in rural England (Riva et al. 2009, Riva et al. 2011), to present an analysis of data from the Health Survey for England (HSE), linked to information on small geographic areas, characterized by their urban/rural settlement profile and socioeconomic conditions.

Data and Analyses

The HSE is an annual cross-sectional survey, representative of the English population (Prior et al. 2003). Health and socio-demographic information on adults aged 18 years and older were pooled for the years 2000-2003. Two psychosocial constructs of health and wellbeing were selected for this research: good self-rated health and absence of psychological distress. Self-rated health is a measure of general health that is predictive of morbidity and mortality (Idler and Benyamini 1997) so it bears an established relationship to presence of medically recognized disease, as well as perceived illness. Responses to the self-rated health question were dichotomized by assigning the value 1 to those reporting their health as being 'very good' or 'good' and 0 to those reporting 'fair', 'bad' or 'very bad' health. Anxiety and depression represent the most common forms of psychological health problems, and these are assessed in the HSE using the 12-item General Health

Questionnaire (GHQ) which has been widely validated against standardized clinical interviews. Positive mental health was dichotomized by assigning the value 1 to those having a score lower than 3 on the GHQ-scale (Goldberg and Williams 1988, Weich et al. 2006) and 0 for those with scores of 3 and above who are likely to be suffering psychological distress. These then are measures of 'good' health or, at least, absence of distress. While being in good health is not synonymous with positive wellbeing, we have argued above that it is one of the conditions that relate to wellbeing. Also, as these are measures of self-reported health status, they seem likely to reflect, to some extent, respondents' individual 'sense of wellbeing'.

Several individual socio-demographic characteristics were considered as these may be associated with health according to previous research: age (in 10-year age groups) and sex; self-reported ethnic group (being British, Scottish, Irish or Welsh vs. reporting membership of another ethnic group); marital status (being single, separated/divorced, widowed versus being in a relationship or cohabiting); household income (adjusted for the number of persons in the household), work status (being in employment versus unemployed, retired, or other economically inactive); occupational social class; having access to a car (which is partly a measure of disposable wealth, but is also especially significant in rural areas for access to work, services and amenities); social support from family and friends; and length of residence in the local area.

Small areas were defined using MSOAs which are aggregates of 'Output Areas', the latter being the smallest units for which census data are published (Office for National Statistics 2006). Compared to electoral wards, MSOAs are more consistent in geographical extent and population size (between 5,000 and 7,200 people) and are designed to maximize homogeneity in terms of housing tenure and dwelling type, to be delimited by obvious boundaries such as major roads, and to have relatively stable boundaries over time (Office for National Statistics 2006). MSOAs were characterized in terms of rurality, socioeconomic deprivation and social cohesion.

The analysis uses a set of rural/urban categories developed by the Department for Environment, Food, and Rural Affairs 2001 to classify small areas (Bibby and Shepherd 2001). Urban areas were classified either as located within the 'Greater London area' or in 'other cities'. This distinction was made because London has distinctive social and economic conditions which distinguish it from 'other cities' in England in ways that may be important for health and wellbeing. Rural areas were classified according to their settlement type, distinguishing between 'small towns' (including towns and fringe settlements) and 'villages' (including villages, hamlets, and isolated dwellings).

Socioeconomic deprivation of MSOAs was measured using the overall score of the 2004 Index of Multiple Deprivation (Noble et al. 2004) which combines seven separate dimensions, or domains, of deprivation experienced by individuals living in an area relating to income, education, employment, health, housing and services, crime and living environment.

Social cohesion of areas was assessed using a social fragmentation index (Congdon 1996) which has been shown to relate to psychological health outcomes independently of measures of material/economic deprivation (Fagg et al. 2008). This indicator is composed of four variables from the 2001 Census: proportion of the population living alone; proportion of adults not in married couples; proportion of residents who moved into the LAD in the last year; and proportion of residents living in private rented accommodation (Congdon 1996). To protect confidentiality of the HSE respondents, MSOA deprivation and social cohesion data were classified into deciles and linked to the anonymized individual survey records before they were released to us by the UK National Centre for Social Research. Because of small numbers of respondents of MSOAs in some deciles, the areas were aggregated for analysis into lower (1 to 4), middle (5 to 6) and higher (7 to 10) categories of deprivation.

To account for the hierarchically structured nature of the dataset, i.e. people 'nested' within areas, multilevel models for dichotomous outcomes were used (Raudenbush and Bryk 2002, Raudenbush et al. 2004) (analyses were conducted for all those in areas with minimum sample of five HSE respondents per MSOA). Multilevel modelling is an approach to regression analysis where the variation in the outcome variable is explained simultaneously by variables measured at the individual-level (composition effects) and at the area-level (contextual effects) (Duncan and Jones 1995, Diez-Roux 2000). These models further allow examination of 'cross-level' interaction, showing whether area effect on health is modified by individuals' characteristics (testing whether for some people, with particular characteristics, health is especially 'sensitive' to variation in local area conditions) and to account for the inter-dependence of observations within groups. Results are presented using odds ratios (OR) and 95 per cent confidence intervals (95 per cent Confidence Interval- CI). OR greater than 1 indicates that the odds of reporting good psychosocial health are greater than in the reference group. Conversely, a value less than 1 indicates a lower likelihood of the outcome than in the reference group. If the 95 per cent CI for this OR does not include the value of 1, then likelihood of the outcome is significantly different, in statistical terms, than in the referent category (the difference is unlikely to have occurred by chance).

Results

Urban-Rural Health Differences

For this first set of results, analyses were based on a sample of 33,520 respondents living in one of 2642 MSOAs. Health status of respondents living in the Greater London area, in semi-rural areas, and in villages was compared to health of those living in 'other cities'. Odds ratios and 95 per cent confidence intervals are reported graphically in Figure 4.2. Overall, results showed that residents of small towns and villages were more likely to report more favourable health status than

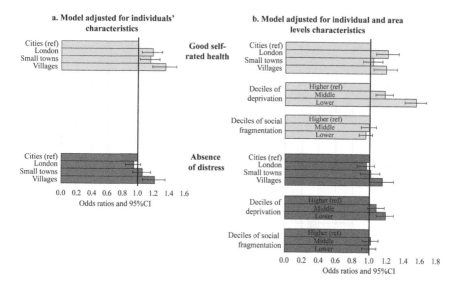

Figure 4.2   **Health outcomes in urban and rural areas, adjusting for individuals' socio-demographic characteristics and area conditions**

residents in other cities. Adjusting the statistical model for the characteristics of individuals that might explain these urban-rural disparities (Figure 4.2a), living in small towns was associated with better self-reported health (OR: 1.15; 95 per cent CI: 1.04, 1.29) whereas living in villages was associated with both better self-rated health (OR: 1.35; 95 per cent CI: 1.20, 1.51) and absence of psychological distress (OR: 1.22; 95 per cent CI: 1.08, 1.37) in comparison to living in urban areas in 'other cities'. The independent health effect of rurality persisted when area deprivation and social fragmentation were considered, as illustrated in Figure 4.2(b): the overall health advantage of rural areas remained statistically significant for villages only, where better psychosocial health outcomes were not explained by people's socio-demographic circumstances or by area-level deprivation and social fragmentation.

Geographic health inequalities were also observed between different types of urban areas, with residents of Greater London being more likely to report better health status than their urban counterparts living in 'other cities', independently of their socio-demographic characteristics. However, they appeared less likely to enjoy good mental health, although this geographic inequality was explained by different area composition and context. In addition to the positive rural health effect, lower levels of area-deprivation were significantly associated with better psychosocial health, although social fragmentation was not.

As reported in other studies relying on UK cohorts (Weich et al. 2006) and other cross-sectional data on mental health (Paykel et al. 2000), results of these analyses are therefore indicative of higher levels of psychological 'wellbeing' in

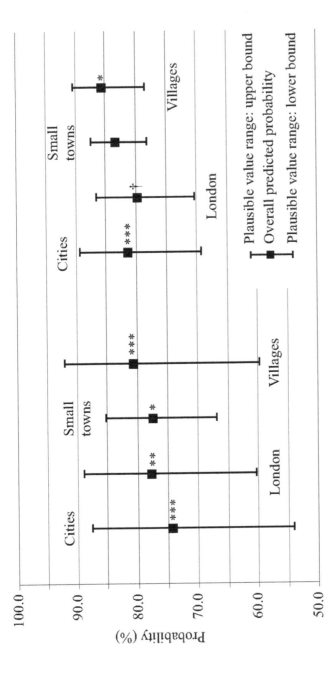

Figure 4.3 **Overall predicted probabilities (and plausible value range) of observing good self-rated health and absence of distress in MSOAs within types of urban and rural areas**

Note: † p < 0.10; *<0.05; ** p< 0.01; *** p<0.001.

Source: Adapted from Riva et al., 2009. *Social Science & Medicine* 68, 654-63.

rural settings in England, expressed through better self-reported 'health' outcomes. With respect to mental health, results may seem to run counter to other research showing higher prevalence of serious psychiatric morbidity (especially mortality from suicide) in remote rural than in suburban areas (Levin and Leyland 2005, Middleton et al. 2006). This may be because factors determining inequalities in the absence of common mental disorders (as examined here) are likely to be rather different from those influencing suicide.

Potential explanations for more favourable psychosocial health outcomes among rural dwellers in England could relate to the absence, in rural places, of environmental risk factors more characteristic of urban areas (Haynes and Gale 1999), including stressful social and material conditions. Controlling for area composition and context attenuated the urban-rural disparities in health, suggesting that health differences may be largely explained by differences in socioeconomic conditions between poor urban localities and more affluent rural places. But, as we discuss next, the overall health advantage of rural areas masks more local inequalities in health.

Intra-rural and Intra-urban Variation in Health

To further investigate geographic inequalities in health, the extent of intra-rural and intra-urban variation in health was examined for respondents grouped into MSOAs in: small towns (208 MSOAs), villages (238 MSOAs), Greater London area (354 MSOAs) and 'other cities' (1842 MSOAs). The aim of these analyses was to compare individual and area inequalities in health occurring within categories of rural settlements to inequalities occurring among urban settlements. Figure 4.3 presents the overall predicted probabilities of observing good self-rated health and positive mental health in the four settlement types; the plausible value range describes the likely variability at MSOA level within these categories (this variation is not adjusted for individuals' characteristics or area conditions) (Raudenbush and Bryk 2002).

Results demonstrate significant geographic health inequalities among small rural areas and among urban settings. In the HSE, on average, about 80 per cent of people living in villages reported good self-rated health, yet this proportion varied significantly: in some villages more than 90 per cent of the respondent reported good health, whereas in other villages less than 60 per cent perceived their health as being good. Significant variation in self-rated health was also observed across small towns although the range of the variation was smaller. The probability of being free of psychological distress also varied significantly for respondents living in villages, but not for those living in small towns. There was also important local variation in the reporting of these two health outcomes across cities and among small areas located in Greater London.

Adjusting for the socio-demographic characteristics of the HSE respondents and for area-level deprivation and social cohesion explained, to some extent, the intra-rural and intra-urban health disparities. Significant geographic and

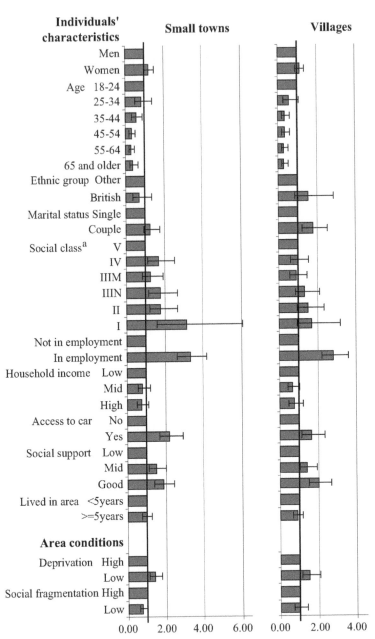

a. I Professional; II Managerial; IIIN: Skilled non-manual; IIIM: Skilled manual;
IV: Semi-skilled manual; V: Unskilled Manual

**Figure 4.4   Individuals' characteristics and area-level conditions associated
with better self-reported health in rural areas of England**

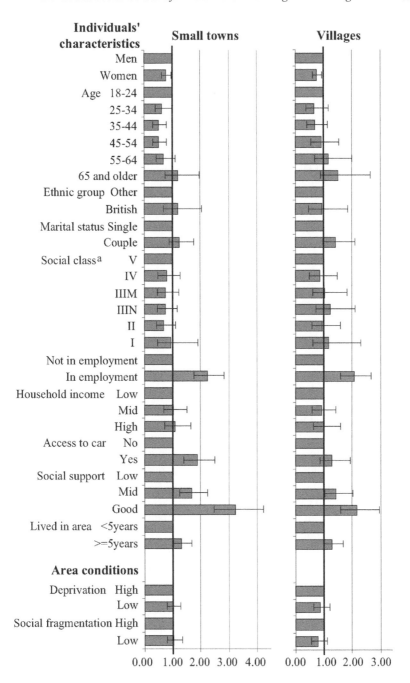

Figure 4.5 **Individuals' characteristics and area-level conditions associated with absence of psychological distress in rural areas of England**

social inequalities in health were observed among rural and among urban areas. Results are presented for rural areas only since we are especially interested in the disparities within rural regions; Figure 4.4 presents shows for good self-rated health and Figure 4.5 for absence of distress.

In rural areas, results generally suggest that better self-reported health is less prevalent among older age groups, and more prevalent among women, among people in work, in higher social class, and those having access to a car (which probably represents better access to facilities and social contacts), among people enjoying better social support and among those living in a less deprived area. Absence of psychological distress was also more likely for those with strong social support and people in work, but otherwise followed a different pattern, being generally less likely for women, older people, and for those who had lived longer in the area. The differing patterns for these two health outcomes demonstrate the complex and multi-dimensional nature of health and wellbeing in populations. Self-reported health is likely to reflect both physical and mental health and wellbeing, while the GHQ focuses more on psychological distress, which may partly explain the different patterns of variation observed. If so, we might speculate, for example, on whether the onset of physical health problems as one grows older may be partly compensated by declining risks of psychological distress. This highlights the value of a life course perspective on health and wellbeing to identify how these aspects of human experience vary over time.

More Local Social Inequalities in Health in Rural Areas

We have shown that being in work and having a car, and being in a higher social group are generally associated with better self-reported health and lower risk of psychological distress. As mentioned above, some authors have argued that the heterogeneous socioeconomic profile of rural populations, within small areas, makes it difficult to establish the significance of rural poverty for health (Haynes and Gale 2000). To explore this issue, a further set of analyses examined whether being in employment is more 'protective' for a person's health outcomes in rural areas or in urban areas.

For this analysis, MSOAs were classified using the 'employment deprivation domain' of the IMD, which indicates the level of economic inactivity at the area-level and conceptualized as 'the involuntary exclusion of the working age population from the world of work' (Noble et al. 2004). Here the study sample is limited to HSE respondents of working age (18 to 64 years old) living in either urban or rural areas (the different types of urban and rural settlements were collapsed in two categories) (results are not tabulated). The benefits for a person's self-rated health associated with being in employment were greater for rural dwellers than those living in cities (OR: 1.33: 95%CI: 1.11-1.31) and risk of psychological distress was also more strongly linked to employment status in rural rather than in urban areas (OR: 1.25: 95%CI: 1.03-1.51). The health benefits associated with being in work were also more important for people residing in areas characterized by higher levels of 'employment deprivation' (self-rated health OR: 1.58; 95%CI: 1.34-1.86; mental health OR: 1.42; 95%CI: 1.21-1.68) than for those in work but living in less deprived areas.

These results emphasize how individual and area factors interact in their relationship with health and wellbeing. Living in rural areas, but also in more deprived urban areas, where employment opportunities might be more limited may be especially damaging to good health and wellbeing for people with more precarious employment status. Also our findings show that socio-economic inequalities in health (associated with employment in this instance) appear as important in rural settings as they are in urban locations.

Being in employment may be important for health because it provides more material resources to individuals, e.g. disposable income, and sense of purpose and meaning associated with work, and protection from stressors in the local environment. A more psychosocial explanation would further suggest that individuals who are in work, and living in communities with a depressed labour market, compare themselves particularly positively to their neighbours excluded from the labour force, and this has a beneficial influence on their mental health. This would support the idea that our sense of health and wellbeing is influenced by the comparisons that we make with those around us and that we are more likely to 'count our blessings' and feel better about our own situation if it compares well with many of the people that we know in our local community. However, we must also bear in mind that the cross-sectional nature of the data means that an alternative interpretation is that in rural areas (as well as the most deprived urban areas nationally) people with health problems are more likely to become unemployed, which would be consistent with greater precarity of employment in economically marginal areas.

Conclusion

This discussion has focused on analysis of 'proxy' measures of wellbeing which are actually more directly related to health, as perceived by the individual. This raises a debate over the relationship between wellbeing and health, which are probably related, but are certainly not synonymous. Here we take the view that better health is an important, but not a necessary condition for sense of 'wellbeing'. Our focus has mainly been on populations of rural areas, which, in many accounts of British society, are presented as having relatively good levels of health and wellbeing. Access to 'therapeutic' 'green' spaces, lower levels of social disorder and stronger traditional social support systems are invoked as likely to contribute to better wellbeing, as well as absence of the intense concentrations of poverty found in some parts of cities.

Our analysis above shows that this notion of the 'rural idyll' may be an over-simplification, as is apparent if one considers very local variability between different parts of rural regions, or differences in individual experiences in rural settings. For example, some parts of rural regions are significantly worse off than others on the measures used. Also some individual circumstances, such as those associated with being in work, are at least as important for health in rural areas as they are in the most deprived parts of cities. Thus future research on the health/ wellbeing complex should examine interactions between individuals and the community settings where they live, in order to better understand the geography of wellbeing in rural areas. We have also suggested that life stage and length of residence in rural areas seem to relate to the health outcomes studied here in ways that suggest a strong case for a 'lifecourse' approach to the study of wellbeing as a potentially fruitful way forward in future research on wellbeing.

Acknowledgements

We gratefully acknowledge the National Centre for Social Research (NatCen) for permitting the use of the Health Survey for England (HSE) for the years 2000 to 2003. Data for HSE 2000-2003 are from the National Centre for Social Research and University College London, Department of Epidemiology and Public Health, Colchester, Essex and distributed by the UK Data Archive. Crown copyright material is reproduced with the permission of the Controller of HMSO and the Queen's Printer for Scotland. Help provided by the staff of NatCen is also thankfully acknowledged. The authors alone are responsible for the interpretation of the data.

References

Australian Institute of Health and Welfare. 2007. *Rural, Regional and Remote Health: A Study on Mortality* (2nd edition). Canberra: Australian Institute of Health and Welfare.

Berkman, L.F. and Kawachi, I. 2000. *Social Epidemiology*. New York: Oxford University Press.

Bibby, P. and Shepherd, J. 2001. *Developing a New Classification of Urban and Rural Areas for Policy Purposes – The Methodology*. London: Department for Environment, Food, and Rural Affairs.

Boland, M., Staines, A., Fitzpatrick, P. and Scallan, E. 2005. Urban-rural variation in mortality and hospital admission rates for unintentional injury in Ireland. *Injury Prevention*, 11, 38-42.

Chaix, B. 2009. Geographic life environments and coronary heart disease: A literature review, theoretical contributions, methodological updates, and a research agenda. *Annual Review of Public Health*, 30, 81-105.

Commission for Rural Communities. 2008. *State of the Countryside Update: Rural Analysis of the Index of Multiple Deprivation 2007*. Cheltenham: Commission for Rural Communities.

Commission for Rural Communities. 2010. *State of the Countryside 2010*. Cheltenham: Commission for Rural Communities.

Congdon, P. 1996. Suicide and parasuicide in London: A Small-area study. *Urban Studies*, 33, 137-58.

Curtis, S. 2004. *Health Inequality: Geographical Perspectives*. London: Sage.

Curtis, S. 2010. *Space, Place and Mental Health*. Farnham: Ashgate.

Curtis, S. and Jones, I.R. 1998. Is there a place for geography in the analysis of health inequality? *Sociology of Health & Illness*, 20(5), 645-72.

Davey-Smith, G., Shipley, M.J. and Rose, G. 1990. Magnitude and causes of socioeconomic differentials in mortality – further evidence from the Whitehall study. *Journal of Epidemiology and Community Health*, 44, 265-70.

Diez-Roux, A.V. 2000. Multilevel analysis in public health research. *Annual Review of Public Health*, 21, 171-92.

Duncan, C. and Jones, J. 1995. Individuals and their ecologies: Analysing the geography of chronic illness within a multi-level modelling framework. *Health and Place*, 1, 27-40.

Eberhardt, M.S. and Pamuk, E.R. 2004. The importance of place of residence: Examining health in rural and nonrural areas. *American Journal of Public Health*, 94, 1682-86.

Fagg, J., Curtis, S., Stansfield, S.A., Cattell, V., Tupuola, A.M. and Arephin, M. 2008. Area social fragmentation, social support for individuals and psychosocial health in young adults: Evidence from a national survey in England. *Social Science & Medicine*, 66, 242-54.

Fernandez-Ballesteros, R., Zamarron, M.D and Ruiz, M.A. 2001. The contribution of socio-demographic and psychosocial factors to life satisfaction. *Ageing and Society*, 21, 25-43.

Fleuret, S. and Atkinson, S. 2007. Wellbeing, health and geography: A critical review and research agenda. *New Zealand Geographer*, 63, 106-18.

Frohlich, K.L. and Potvin, L. 2008. The inequality paradox: The population approach and vulnerable populations. *American Journal of Public Health*, 98, 216-21.

Gartner, A., Farewell, D., Dunstan, F. and Gordon, E. 2008. Differences in mortality between rural and urban areas in England and Wales, 2002-04. *Health Stat Q*, 39, 6-13.

Goldberg, D.P. and Williams, P. 1988. *The User's Guide to the General Health Questionnaire*.

Hartley, D. 2004. Rural health disparities, population health, and rural culture. *American Journal of Public Health*, 94(10), 1675-78.

Haynes, R. and Gale, S. 1999. Mortality, long-term illness and deprivation in rural and metropolitan wards of England and Wales. *Health & Place*, 5(4), 301-12.

Haynes, R. and Gale, S. 2000. Deprivation and poor health in rural areas: Inequalities hidden by averages. *Health & Place*, 6(4), 275-85.

Idler, E. and Benyamini, Y. 1997. Self-rated health and mortality: A review of twenty-seven community studies. *Journal of Health and Social Behavior*, 38, 21-37.

Jones, K. and Moon, G. 1993. Medical geography: Taking space seriously. *Progress in Human Geography*, 17, 515-24.

Kearns, R.A. 1993. Place and health: Towards a reformed medical geography. *Professional Geographer*, 45, 139-47.

Kearns, R.A. and Joseph, A.E. 1993. Space in its place: Developing the link in medical geography. *Social Science and Medicine*, 37, 711-17.

Kelleher, C.C., Friel, S., NicGabhainn, S. and Tay, B. 2003. Socio-demographic predictors of self-rated health in the Republic of Ireland: Findings from the National Survey on Lifestyle, Attitudes and Nutrition, SLAN. *Social Science & Medicine*, 57, 477-86.

Kim, D. 2008. Blues from the Neighborhood? Neighborhood Characteristics and Depression. *Epidemiologic Reviews*, 30(1), 101-17.

Kyte, L. and Wells, C. 2010. Variations in life expectancy between rural and urban areas of England, 2001-07. *Health Stat Q*, 46, 25-50.

Levin, K.A. 2003. Urban-rural differences in self-reported limiting long-term illness in Scotland. *Journal of Public Health Medicine*, 25(4), 295-302.

Levin, K.A. and Leyland, A.H. 2005. Urban/rural inequalities in suicide in Scotland, 1981-1999. *Social Science & Medicine*, 60(12), 2877-90.

Lynch, J.W., Kaplan, G.A. and Salonen, J.T. 1997. Why do poor people behave poorly? Variation in adult health behaviours and psychosocial characteristics by stages of the socioeconomic lifecourse. *Social Science & Medicine*, 44, 809-19.

Macintyre, S., Ellaway, A. and Cummins, S. 2002. Place effects on health: How can we conceptualise, operationalise and measure them? *Social Science & Medicine*, 55, 125-39.

Macintyre, S., Maciver, S. and Sooman, A. 1993. Area, class and health: Should we be focusing on places or people? *Journal of Social Policy*, 22, 213-34.

Mair, C., Diez-Roux, A.V. and Galea, S. 2008. Are neighbourhood characteristics associated with depressive symptoms? A review of evidence. *Journal of Epidemiology and Community Health*, 62(11), 940-46.

Middleton, N., Sterne, J. A. and Gunnell, D. 2006. The geography of despair among 15-44-year-old men in England and Wales: Putting suicide on the map. *Journal of Epidemiology and Community Health*, 60(12), 1040-47.

Noble, M., Wright, G., Dibben, C., Smith, G.A.N., McLennan, D., Anttila, C., Barnes, H., Mokhtar, C., Noble S., Avenell, D., Gardner, J., Covizzi, I. and Lloyd, M. 2004. *Indices of Deprivation 2004*. London: Neighbourhood Renewal Unit.

Office for National Statistics. 2006. *Beginners' Guide to UK Geography – Super Output Areas (SOAs)* [Online]. Available at: www.ons.gov.uk/about-statistics/geography/products/geog-products-area/names-codes/soa/index.html [Accessed: 25 May 2011].

Pampalon, R., Martinez, J. and Hamel, D. 2006. Does living in rural areas make a difference for health in Quebec? *Health & Place*, 12(4), 421-35.

Panter-Brick, C. and Fuentes, A. 2010. Health, Risk, and Adversity: A Contextual View from Anthropology, in *Health, Risk, and Adversity*, edited by C. Panter-Brick and A. Fuentes. New York: Berghahn Books, 1-10.

Patychuk, D. L. 2007. Bridging place-based research and action for health. *Canadian Journal of Public Health-Revue Canadienne de Santé Publique*, 98, S70-73.

Paykel, E.S., Abbott, R., Jenkins, R. Brugha, T.S. and Meltzer, H. 2000. Urban–rural mental health differences in Great Britain: Findings from the National Morbidity Survey. *Psychological Medicine*, 30(2), 269-80.

Prior, G., Deverill, C., Malbut, K. and Primatesta, P. 2003. *Health Survey for England 2001: Methodology and Documentation*. London: The Stationery Office.

Raudenbush, S. and Bryk, A. 2002. *Hierarchical Linear Models: Applications and Data Analysis Methods*. Newbury Park: Sage.

Raudenbush, S., Bryk, A., Cheong, Y. and Congdon, R.T. 2004. *HLM 6: Hierarchical Linear and Nonlinear Modeling*. Chicago, IL: Scientific Software International.

Riva, M., Bambra, C., Curtis, S. and Gauvin, L. 2011. Collective resources or local social inequalities? Examining the social determinants of mental health in rural areas. *European Journal of Public Health*, 21(2), 197-203.

Riva, M., Curtis, S., Gauvin, L. and Fagg, J. 2009. Unravelling the extent of inequalities in health across urban and rural areas: Evidence from a national sample in England. *Social Science & Medicine*, 68(4), 654-63.

Riva, M., Gauvin, L. and Barnett, T.A. 2007. Toward the next generation of research into small area effects on health: A synthesis of multilevel investigations published since July 1998. *Journal of Epidemiology and Community Health*, 61(10), 853-61.

Stafford, M. and Marmot, M. 2003. Neighbourhood deprivation and health: Does it affect us all equally? *International Journal of Epidemiology*, 32(3), 357-66.

Stratton, A. 2010. Happiness index to gauge Britain's national mood. *The Guardian*. [Online, 14 November] Available at: www.guardian.co.uk/lifeandstyle/2010/nov/14/happiness-index-britain-national-mood [accessed: 3 December 2010].

Weich, S., Twigg, L. and Lewis, G. 2006. Rural/non-rural differences in rates of common mental disorders in Britain – Prospective multilevel cohort study. *British Journal of Psychiatry*, 188, 51-7.

WHO. 1946. *Preamble to the Constitution of the World Health Organization* as adopted by the International Health Conference, New York, 19-22 June, 1946; signed on 22 July 1946 by the representatives of 61 States (Official Records of the World Health Organization, no. 2, 100) and entered into force on 7 April 1948.

WHO.1978. *Declaration of Alma Ata*. International conference on primary health care, Alma-Ata, USSR, 6-12 September 1978. Geneva.

WHO Commission on the Social Determinants of Health. 2008. *Closing the Gap in a Generation: Health Equity through Action on the Social Determinants of Health*. Geneva: World Health Organization.

Wilkinson, R. and Pickett, K.E. 2007. *The Spirit Level: Why more Equal Societies Almost Always do Better*. London: Allen Lane.

Young Foundation. 2008. *Neighbourliness + Empowerment = Wellbeing: Is there a Formula for Happy Communities?* [Online] Available at: www.youngfoundation.org/files/images/N_E_W_web_v4.pdf [accessed: 11 January 2011].

Chapter 5

Wellbeing and the Neighbourhood: Promoting Choice and Independence for all Ages

Rose Gilroy

From Easterlin (1974) onward, research has demonstrated that, once earnings reach a level of comfort and competence, there is no appreciable increase in life satisfaction or happiness. In recent times this finding has moved beyond the level of individual and nation to mesh with holistic concepts of sustainable development that suggest human flourishing, in all its dimensions, has been sacrificed in our drive toward greater growth. In the UK, the influential New Economics Foundation set out a wellbeing manifesto in an attempt to define on what issues central government could take action that would promote individual and society wide wellbeing (Shah and Marks 2004). A range of government departments, including the Department of Environment Food and Rural Affairs (Dolan et al. 2006) and the Department of Trade and Industry (the Foresight programme), have attempted to revise their strategies around wellbeing.

The discussion that follows weaves a third strand into the wellbeing agenda: if an achievement of this century is a marked global increase in longevity, how do we ensure that these extra years are worth living? How can societies promote wellbeing or human flourishing throughout the life course? The established critical contribution of place to supporting people in later life has been given new policy prominence by the WHO's Age Friendly City research (2007) and in the UK by a commitment to creating lifetime neighbourhoods (DCLG 2008). This prompts a discussion that explores the impact of broader societal changes on neighbourhoods and consequently on older people.

A New Landscape of Ageing

Later life, through the lens of role theory (George 1998), has often been discussed as one in which subjective wellbeing is eroded as people are progressively stripped of the major adult roles of life: worker, spouse, and autonomous decision-making individual. The rise of 'active ageing', which is now the dominant discourse, has provided a counter argument that old age can also be marked by acquisition of new roles and, potentially, be a flourishing time of mobility and new creativity. Far from being a time of erosion of identity markers, older people may remake themselves through consumption (Rees Jones et al. 2008) though, clearly, this demands a level of wealth that not all older people are able to command. The dominance of the active ageing agenda is also underpinned by the retirement of the post-World War II baby boomers, a cohort that was at the forefront of the consumer culture and the youth cultures of the 1960s and 1970s. This cohort, retiring with better health, more wealth and more confidence in their power to make changes, can potentially reshape societies through the choices that they make. Where they choose to live, the services they demand, their appetite for continued employment, greater engagement in adventure holidays and their desire to continue or leave employment all have an impact on our labour markets, housing markets and ecological footprint. As a society we need to consider how we may best make use of the resources of time, experience and maturity that this group offers.

The 'have it all' view of later life needs to be tempered, however, with the growth of the oldest old – the over 85s – who, while potentially healthier than their parents' generation, are likely to be faced with physical frailty, cognitive decline and a need for support.

Table 5.1   UK population of those aged 65 years and over projected to 2030 by percentage increase from 2010

	2010	2015	2020	2025	2030
People aged 65-69	0	22	11	21	39
People aged 70-74	0	9	34	22	34
People aged 75-79	0	9	21	49	37
People aged 80-84	0	8	22	37	71
People aged 85-and over	0	15	36	66	101
Total population aged 65 and over	0	13	23	35	51

Source: Projecting Older People Population Information System 2010.

By 2031 the numbers of those aged 85 or older that require support at regular intervals will increase from 215,000 to 384,000 and those who require 24-hour-care will increase from 302,000 to 541,000 (Bond 2010). The contraction in family size and the mobility of families coupled with the demand to ration and contract formal welfare services, raises two major challenges: how will we care for frail older people and who, in turn, will provide services for our future selves?

Policy Responses to Ageing in the UK

A raft of policy changes and documents has appeared in response to the changing age structure and, while the focus has varied, together they form a mosaic that aims to incrementally increase the choices of older people. Making choices is a major plank in independence that is particularly valued by older people (Fisk and Abbott 1998). Some policy changes enlarge the range of everyday choices by increasing individuals' financial resources, including increases in the state pension, proposals for supporting continued economic activity and the introduction of free local bus travel. Choice can be expressed by having outlets for sharing ideas, aspirations, and feedback. A range of local authority services are more active in engaging older people in strategy making while past initiatives such as Better Government for Older People and more recently Communities in Control, Citizens' Assemblies, the National UK Advisory panel (and its face in the regions) work to provide opportunities from local through to national levels for the raising of older people's concerns. Through enhanced dialogue, there has been a raft of policy documents that have recognized older people's perspectives (Gilroy 2003, 2008).

A range of policies promote choice and independence, most often understood as increased support to live at home. Attempts to increase major life choices about housing in later life have been addressed by policy documents ranging from the Department of Health and Department for Environment, Transport and the Regions (2002) *Quality and Choice for Older People's Housing: A Strategic Framework* through to *Lifetime Homes, Lifetime Neighbourhoods: A National Strategy for Housing in an Ageing Society* (DCLG and DWP 2008) and the many discussion papers on topics such as lifetime homes or handyperson schemes. The Ministerial Concordat *Putting People First: Transforming Adult Care* (HM Government 2007) and the Department of Health (2007) *Commissioning Framework for Health and Wellbeing* introduced a concept of personalized budgets that aim to make a radical and sustained shift in the way in which services are delivered, ensuring that they are more personalized. The impact of this policy thrust needs to be considered in the light of the likely increase in the number of over 85s who are living alone and pressures on funding social care. If families are no longer able to offer much support and local government and voluntary sector services are strained then older people may have to rely increasingly on the physical and social infrastructure of their neighbourhood. As the WHO age friendly city research (2007), has emphasized it is not only the accessibility and quality of the physical place, but also the more intangible qualities of respect and social inclusion that promote choice making by older people. The discussion turns now to consider the concept of everyday life as a lens through which to consider the adequacy of neighbourhood.

Taking an Everyday Life Perspective on Wellbeing

The concept of everyday life (set out in Figure 5.1) is an understanding of life as a web of social relations through which we accomplish human existence in daily, weekly, yearly, lifespan and inter-generational time (Healey 1997). It explores the task of acquiring the means of financial competence; of social support which may be emotional, spiritual, and practical; the necessity of pleasure and play in our lives; the validation of an individual by having arenas for citizenship namely the opportunity to speak and be heard (Gilroy and Booth 1999).

making ends meet

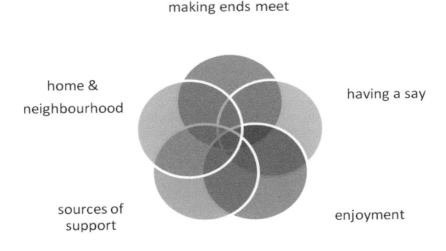

home &
neighbourhood

having a say

sources of
support

enjoyment

Figure 5.1   The Everyday Life Model

Critical to an understanding of the role played by place in supporting wellbeing, the everyday life approach explores home and neighbourhood as both spatial and social envelopes. It emphasizes both the individual in their home and on the street but also it encapsulates a being together in shared space; a connection through common activities performed outside and a spatial container for shared action. This meshes with a human ecology approach that acknowledges the centrality of environment and breaks this down into social and physical components seeing people as both shapers of and shaped by the environments in which they are embedded. These social and physical environments are permeable with changes in one brought about by individual actors or through policy mechanisms interacting with and influencing others (Keating and Phillips 2008).

While this concept originated in the Nordic Feminist Network (see Horelli and Vepsa 1994 for this discussion), its applicability to later life is clear. People aged over 65 are estimated to spend more than 85 per cent of their time in the home and

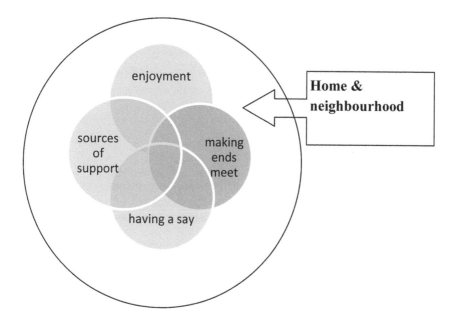

Figure 5.2   The Geography of Everyday Life

this rises to 90 per cent for those aged 85 or older (House of Lords Science and Technology Committee 2005). There may be many reasons for staying at home including losing chosen companions, falling income and decreasing health but research also consistently reveals a greater attachment to home in older people. The seminal work of Belk (1988) explores the role of possessions and mementoes as mirrors of identity. Clare Cooper Marcus (1995) talks of the 'emotional nurturance' that older people may find in their home. This attachment – what Rowles (1983) has called 'autobiographical insideness' – can flow beyond the home into the locality (Peace et al. 2006). Research reveals older people, particularly women, acting as 'neighbourhood keepers' (Phillipson et al. 1999: 741).

Janevic, Gjonca and Hyde have argued that as mobility decreases older people may 'depend to an even greater extent than younger people on what their local environment and personal social networks have to offer' (2003: 313). The small and the local may increasingly be the arenas in which older people must meet their range of everyday needs:

> Concepts such as successful ageing, ageing in place and quality of life have informed many of the national strategies developed to date to address issues associated with ageing populations. The promotion of physical health, economic independence and social connectivity are key features of such community activity beyond the immediate home environment. This also implies an urban

environment capable of enabling the broadest range of functional limitations in a sustainable way. (Landorf et al. 2008: 512)

Therefore, it may be appropriate to modify the everyday life model and to see the home and neighbourhood as the spatial and social arenas in which older people need to meet their everyday life needs (see Figure 5.2).

Making Ends Meet: Tough Choices

Though making ends meet also refers to the availability of employment, for older people beyond the waged work phase of life it is concerned with managing within an income. A bleak report from the New Economics Foundation in 2003 reported on the haemorrhaging of neighbourhood facilities:

> Between 1995 and 2000, the UK lost 20 per cent of some of its most vital institutions: corner shops, grocers, high street banks, post offices and pubs, amounting to a cumulative loss of over 30,000 local economic outlets. (Oram et al. 2003: 2)

The impact on neighbourhoods of the current recession is unclear, though Batty and Cole (2010) reporting on deprived areas give vivid examples of the closure of the last shop in a parade of empty retail units. For those, like many elders, who may have limited mobility, this can be a severe loss. The expansion of supermarkets may present greater opportunities for savings and the rise in price of food staples (Midgely 2009) has created greater competition between them but their better value is often tied to "offers" such as `Buy One Get One Free` targeted at families. For a single person on a fixed income these may be unaffordable and, for a frailer older person, the weight of goods may too heavy to carry. This argument is also predicated on good accessibility. Supermarkets, particularly those offering clothes and electrical items are located close to major road networks that advantage those with cars or access to them. Forty five per cent of those aged 70 or over report a mobility problem that makes travel difficult (Age Concern and Help The Aged 2009). Only 31 per cent of over 65s living on their own have access to a car. At the same time that over 60s gained the benefit of free local bus travel, many bus routes were discontinued leaving older people reliant on expensive taxis or community transport services that are neither free nor secured.

The advantages to those with limited mobility of shopping by internet fail to benefit those most in need as only 18 per cent of people aged 65 or older have ever used the internet (Help the Aged 2008). A survey undertaken by MORI indicates that the main deterrent to internet access is not skill or confidence but affordability (Ofcom 2009). Further research needs to understand what impact this is currently having on quality of life for older people given that more services are moving to remote access which promotes convenience (for some) at the expense of human

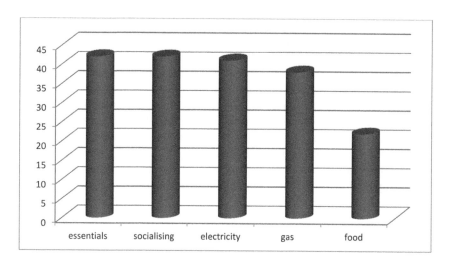

Figure 5.3   Percentage of people aged 60 and over making economies by categories of everyday expenditure

Source: Age Concern and Help the Aged 2009.

interaction that, for many with a shrinking circle of friends and family, may be far more critical.

A 2009 survey of people aged 60 and over paints a bleak picture of personal economy measures that potentially impact on health and wellbeing (Age Concern and Help the Aged 2009).One of the most striking economies made is in domestic fuel to reduce the drain on resources from high heating bills. The 19 per cent of pensioner households living in poverty and the many more on fixed incomes are more sensitive to increasing cost of everyday basics. The RPI sub price index rose for gas by 31 per cent and for electricity by 24 per cent between February 2007 and 2009 alone (Leicester et al. 2009). The subsequent decrease in fuel prices has not been sustained and continued global concerns about oil and gas reserves mean that home heating is likely to take an increasingly larger bite out of household budgets.

The misery of fuel poverty from living in a property that cannot be heated to a comfortable level is a reality for 1.72 million pensioner households – an increase of 250,000 households since 2009 (Department of Energy and Climate Change 2010). Moreover, older people dominate those living in fuel poverty, reflecting in part that more than third of all households living in sub-standard accommodation are headed by an older person (Age UK 2009). Property based measures such as improved insulation and heating packages have improved the comfort levels of many households but older people are least likely to benefit from these because they are disproportionally affected by reduced income, under occupation of properties and, in single person households, have sole responsibility for fuel bills. Preston, Moore and Guertler (2008) argue that this group, that is so frequently affected by poverty, would gain most in personal comfort *and* contribute most to the reduction of carbon emissions by moving to new 'fit-for-purpose' accommodation.

Findings from the *Housing our Ageing Population Panel for Innovation* (HCA 2010) discuss spacious, attractive, and well located housing for older people which, in continental Europe, encourages many older people to move from their family home. This is in contrast to the often cramped and institutional looking specialist housing that older people are offered in the UK. There is an imperative for the UK residential sector to develop better models that meet the needs and aspirations of current and future generations of retirees. What is interpreted as an overwhelming wish to remain in the family home may, in fact, reflect a paucity of high quality options as much as strength of attachment.

Enjoyment: Retaining Vibrancy in the Face of Change

Research in the late nineties in former Welsh mining communities (Drake 1999) revealed the consequences for older people of the loss of vibrancy. The closure of post offices, fish and chip shops, hairdressers, snooker halls, leisure centres, swimming baths and general dealers had created a community of older people who often had to look to the voluntary sector for welfare services. Deficiencies of place were too easily re-interpreted as problems related to reduced mobility and frailty and thus situated within older people themselves. For people whose daily round is confined to smaller arenas this thinning of neighbourhood facilities is yet another weakening of social contact.

Allen (2008) has talked about the impact of isolation and how social contact can be a 'magic lever'. Efforts to increase social contact in the form of access to learning, community and exercise groups can be enormously cost effective. A sense of emotional wellbeing reduces pressure on admissions to care homes but also, if not more importantly, brings older people into the heart of their communities where they can make a contribution. The need for economies in local authority budgets and consequently also on provision by the voluntary sector has led to the closure of day centres and lunch clubs that provided much needed social interaction as well as opportunities for enjoyable activities. There is a real need for the impact on older people's health and wellbeing to be considered before cutting local services and working to ensure that credible alternatives are in place (Age UK 2009). The emphasis on education and skills to return people to the job market should be balanced by the positive impact on health and self-esteem to be gained by older people who have life-long learning opportunities.

Communities are constituted in many ways that are not spatial and, among these, faith groups provide examples of how other important societal structures continue to provide activities for older members:

> On Monday I go to the leisure club. Tuesday there is a lunch. On Wednesday I
> go to the drop in and on Thursday I go to the Jewish home and play cards. Friday

is Shabbat so I go to Shule [synagogue] and the week is gone. In between I go shopping and see my friends, I read and study with the Rabbi. (Mrs M, quoted in Gilroy 2009: 21)

But this is not universal:

We Sikhs spend a lot of time at the Gurdwara. There are women's groups there and we cook together. Hindus go to the temple on Sunday and Muslim ladies go to the Mosque on Fridays but there are no activities for women. (Asian focus group quoted in Gilroy 2009: 21)

And may reflect relative wealth of social capital. With greater support, some groups might increase their capacity for action.

Sources of Support: What is really out there?

The majority of older people frequently mention good social relationships as critical in quality of life. Research demonstrates that more than 60 per cent of older people see family and friends at least once a week either by making visits or being visited, though those over 75 are more dependent on the latter (Bowling et al. 2002). Creating safe neighbourhoods with safe, accessible and reliable transport underpin the maintenance of good social relations.

For those whose lives are limited by disability, support in the form of home delivered care services may be critical. Even before the current financial crisis many local authorities were rationing access by tightening the eligibility criteria so that the majority of older people whose needs are for low-level support are ignored in favour of the minority whose needs are acute. While there are short term financial advantages in this strategy, there are proven financial and societal benefits to increasing low-level support (Personal Social Services Research Unit 2010) that enables individuals to maintain an independent life. The contraction of services for older people has also impacted on sheltered housing, which attracts many older people precisely because it offers an umbrella of support. The retreat from on-site residential wardens to daily phone calls and occasional visits may cause some residents to make the move to residential care while other potential beneficiaries of sheltered housing may question what value it now offers (McCormick 2009). The impact of service contraction is to leave many older people struggling to maintain independence.

The recently formed coalition government in the UK offers a new approach to local services in the form of the 'Big Society' that seeks to transform the relationship between individuals, the state and the community and voluntary sector. While the detail has yet to emerge this seems to offer a range of new possibilities for older people. There may be new opportunities for benefiting from a new landscape of voluntary sector and community organizations working in small places to identify and deliver support and services. As Coote asserts, neighbours working together can be a catalyst for many positive changes:

> When individuals and groups get together in their neighbourhoods, get to know each other, work together and help each other, there are usually lasting benefits for everyone involved: networks and groups grow stronger, so that people who belong to them tend to feel less isolated, more secure, more powerful and happier. (Coote 2010: 2)

The contribution of older people as carers may be also properly evaluated and celebrated. In 2006, nearly 48,000 people aged 85 and over provided unpaid care to another family member or friend (ONS 2009). Wheelock and Jones (2002) have discussed the emotional benefit to all family members from grandparent care and have called for a grandparent allowance that acknowledges their contribution to productivity. At a time when older people may need support themselves, the concept of inter-dependence is central to their concept of a good old age, with themselves as both receivers *and providers* of emotional, practical and social support (Godfrey, Townsend and Denby 2004).

The extent to which older people can play an active role in this Big Society is uncertain. Research by McCormick (2009) on expectations of later life reveals both positive and discouraging attitudes. Respondents asserted that that the peak age for contributing most to society in non-financial ways was in the second half of life with more than one-third stating that people aged over 55 contribute most. This rose to almost half for respondents who themselves were aged over 55. Worryingly less than a quarter of the under-25s agreed with this view and overall only 1 per cent thought that people 75 and over contribute most. Challenging assumptions about old age and finding valued roles for frailer elders is surely one of the many challenges thrown up by constructing a Big Society with an ageing population.

Having a Say

Analysis of voter behaviour suggest that the exercise of democratic rights increases with age such that 75 per cent of those aged 65 and over cast their vote compared to only 37 per cent of those under the age of 25 (Davidson 2008). While it is difficult to say there is a single agenda that binds older voters, nevertheless, their greater willingness to vote and their dominance in an increasing number of UK constituencies, increases the importance of their concerns. That older people are increasingly minded to make their voice heard, can be seen in reports of older people willing to face prison rather than pay rises in their council tax (Hetherington and Wintour 2004) as well as the dogged determination of rank and file activists to tackle national issues:

I am thinking of things that are in front of you that you can maybe take on or do something about for example joining the pensioners association We campaign for better pensions, linked to wages so even if we were on a par with the lowest wage or minimum wage at least when they got an increase we would get one too. (Member of North East Pensioners Association, quoted in Gilroy 2009: 34)

At the local level the 'Speaking Up for Our Age' programme, run by Help the Aged (now Age UK), is dedicated to supporting older people's forums providing opportunities for local campaigning and, through involvement, working to eradicate isolation. There are more than 600 of these groups with a total membership of 200,000 older people. A brief glance at these reveals older people talking about their contribution to the sustainability agenda (Meridian and Downs); petitioning to save disability benefits (Skegness); and campaigning against care home closures (Highlands). One of these forums, Newcastle Elders Council, works to create local change across the domains of everyday life and has successfully shifted from being an organization to be consulted by service providers, to being an equal partner in local strategy-making, particularly on housing matters. While this is progress, older people in Newcastle call for a culture shift that views them as capable actors who can contribute to policy and places, not only to progress an older person's agenda, but for the benefit of all:

The idea of an elder friendly city is that people have to do things for you – you know put seats in the shops- it doesn't say anything about what older people can do about quality of life in the city by coming together and finding things they can take charge of ... Being able to have a voice – it's about wisdom, different sets of values, being disconnected from all the consumerist pressures. That democratic element of giving support by being present, giving support by giving their voices is a very important aspect of democracy. (Mr F. quoted in Gilroy 2009: 35)

Conclusion

It has long been understood that the environmental context is critical for older people's continuance of a quality of life (Phillipson 2004, Phillips et al. 2005, WHO 2007). Older people, facing physical or cognitive change, may rely on their environmental resources to preserve their preferred pattern of living. Quality of life research with older people, suggests that their needs can be best supported by enriched places where there is space and time to talk in high quality public space, neighbourhoods and cities with distinctive architecture, decent quality shops, street markets, pocket parks and community based activities. In these austere times, calls for investment to create these age friendly places may seem misplaced. However ensuring maximum use of facilities such as libraries and schools for a variety of community uses; persuading shops and service providers to open up toilets to the public; improving access to up-to-date information on activities and services;

setting up user groups for parks to facilitate inter-generational planning are all measures that can promote wellbeing and cost very little. At the more formal level festivals and street markets can all provide opportunities to be involved, have fun and promote a sense of local identity. New ways to promote local pride and social capital through community gardens have all proved effective. These solutions are well known. However as this discussion has considered, turbulence at the global economic level threatens public, private and voluntary sectors and causes ripples that impact on small places by stripping out shops and services, threatening care and support and reducing opportunities for activities that may be critical in ensuring that frailer older people have any real quality of life in the neighbourhood.

As in all difficult times there are opportunities to be exploited and it may be that the UK enters a new age of collective action based on collaboration and mutual support. Such a society could gain a great deal from recognizing the untapped resource of older people. Such a society might really be building age friendly places.

References

Age Concern and Help the Aged. 2009. *Coping with the Crunch: The Consequences for Older People.* London: Age Concern and Help the Aged.

Age UK. 2009. *One Voice: Shaping our Ageing Society.* London: Age UK.

Allen, J. 2008. *Older People and Wellbeing.* London: Institute of Public Policy Research.

Batty, E. and Cole, I. 2010. Ramifications of recession. *Town and Country Planning*, 79(1), 27-31.

Belk, R.W. 1988. Possessions and the extended self. *Journal of Consumer Research*, 15, 139-68.

Bond, J. 2010. *Is 85 the new 70?* Paper to the seminar 'Healthy Active Ageing'. Newcastle: Newcastle University, 8th October 2010.

Bowling A., Gabriel Z., Banister, D. and Sutton, S. 2002. *Adding Quality to Quality: Older People's Views on their Quality of Life and its Enhancements.* GO Findings 7, Growing Older Programme, Sheffield: University of Sheffield.

Cooper Marcus, C. 1995. *House as a Mirror of Self: Exploring the Deeper Meanings of Home.* Berkeley, CA: Conari Press.

Coote, A. 2010. *Ten Questions about the Big Society and Ten Ways to Make the Best of it.* London: New Economics Foundation.

Davidson, S. 2008. *Quantifying the Changing Age Structure of the British Electorate 2005-2025: Researching the Age Demographics of the New Parliamentary Constituencies.* London: Age Concern.

Department of Communities and Local Government. 2008. *Lifetime Homes, Lifetime Neighbourhoods: A National Strategy for Housing in an Ageing Society.* London: DCLG.

Department of Energy and Climate Change. 2010. *Fuel Poverty Statistics 2010* [Online]. Available at: http://www.decc.gov.uk/assets/decc/Statistics/fuelpoverty/615-pn10-106.pdf [accessed: 16 October 2010].

Department of Health and Department of Environment, Transport and the Regions. 2001. *Quality and Choice for Older People's Housing: A Strategic Framework*. London: DETR.

Department of Health. 2007. *Commissioning Framework for Health and Wellbeing*. London: Department of Health.

Dolan, P., Peasegood, T., Dixon, A., Knight, M., Phillips, D., Tsuchiya, A. and White, M.P. 2006. *Research on the Relationship between Wellbeing and Sustainable Development*. London: DEFRA.

Drake, R. 1999. *Understanding Disability Policies*. London: Macmillan.

Easterlin, R. A. 1974. Does economic growth improve the human lot, in *Nations and Households in Economic Growth: Essays in Honour of Moses Abramovitz*, edited by P.A. David and M.W. Reder. New York: Academic Press, 89-125.

Fisk, M. and Abbott, S. 1998. Older people and the meaning of independence. *Generations Review*, 8(2), 9-11.

George, K. 1998. Self and identity in later life: Protecting and enhancing the self. *Journal of Aging and Identity*, 3, 133-52.

Gilroy, R. 2003. Why can't more people have a say: Learning to work with older people. *Ageing and Society*, 23(5), 659-74.

Gilroy, R. 2008. Places that support human flourishing: Lessons from later life. *Planning Theory and Practice*, 9(2), 145-63.

Gilroy, R. 2009. *Elder Count: Developing an Agenda*. unpublished development report for Joseph Rowntree Foundation.

Gilroy, R. and Booth, C. 1999. Building an infrastructure for everyday life. *European Planning Studies*, 7(3), 307-24.

Godfrey, M., Townsend, J. and Denby, T. 2004. *Building a Good Life for Older People in Local Communities*. York: York Publishing Services for Joseph Rowntree Foundation.

Healey, P. 1997. *Collaborative Planning: Shaping Places in Fragmented Societies*. London: Macmillan.

Help the Aged. 2008. *Isolation and Loneliness*. London: Help the Aged.

Hetherington, P. and Wintour, P. 2004. Spectre of council tax revolt looms, *The Guardian*, 20th February.

HM Government. 2007. *Putting People First: A Shared Vision and Commitment to the Transformation of Adult Social Care*. London: Department of Health.

Horelli, L. and Vepsa, K. 1994. In search of supportive structures for everyday life, in *Women and the Environment, Human Behaviour and Environment*, edited by I. Altman and A. Churchman, 201-26.

House of Lords Science and Technology Committee. 2005 *Ageing: Scientific Aspects*. Volume 1. London. The Stationery Office.

Housing and Communities Agency (HCA). 2010. *HAPPI: Housing our Ageing Population Panel for Innovation* [Online]. Available at: www.

homesandcommunities.co.uk/public/documents/Happi%20Final%20 Report%20-%20031209.pdf. [accessed: 24 June 2010].

Janevic, M., Gjonca, E. and Hyde, M. 2003. Physical and social environment, in *English Longitudinal Study of Ageing*, edited by M. Marmot, J. Banks, R. Blundell, C. Lessof and J. Nazroo. London: Institute for Fiscal Studies, 301-6.

Keating, N. and Phillips, J. 2008. A critical human ecology perspective on rural ageing, in *Rural Ageing: A Good Place to Grow Old*, edited by N. Keating. Bristol: Policy Press, 1-10.

Landorf, C., Brewer, G., Sheppard, L.A. 2008. The urban environment and sustainable ageing: Critical issues and assessment indicators, *Local Environment*, 13(6), 497-514.

Leicester, A., O'Dea, C. and Oldfield, Z. 2009. *The Expenditure Experiences of Older Households*. London: Institute of Fiscal Studies.

McCormick, J., Clifton, J., Sachrajda, A., Cherti, M. and McDowell, E. 2009. *Getting On: Wellbeing in Later Life*. London: Institute of Public Policy Research.

Midgley, J. 2009. *Just Desserts? Securing Global Food Futures*. Newcastle: Institute of Public Policy Research.

Ofcom. 2009. *The Consumer Experience 2009, Research Report*. London: Ofcom.

Office of National Statistics (ONS). 2009. *Health and Social Care* [Online]. Available at: http://www.statistics.gov.uk/cci/nugget.asp?id=1268 [accessed: 28 October 2010].

Oram, J., Conisbee, M. and Simms, A. 2003. *Ghost Town Britain II: Death on the High Street*. London: New Economics Foundation.

Peace, S., Holland, C. and Kellaher, L. 2006. *Environment and Identity in Later Life*. Maidenhead: Open University Press.

Personal Social Services Research Unit for Department of Health. 2010. *National Evaluation of Partnerships for Older People Projects: Final Report*. London: Department of Health.

Phillips, D.R., Siu, O., Yeh, A. and Cheng, K. 2005. Ageing and the urban environment, in *Ageing and Place; Perspectives, Policy, Practice*, edited by G.J. Andrews and D.R. Phillips. New York: Routledge, 147-63.

Phillipson, C., Bernard, M., Phillips, J. and Ogg, J. 1999. Older people's experiences of community life: Patterns of neighbouring in three urban areas. *The Sociological Review*, 47(4), 715-43.

Phillipson, C. 2004. Urbanisation and ageing: Towards a new environmental gerontology. *Ageing and Society*, 24(6), 963-72.

Preston I., Moore, R. and Guertler, P. 2008. *How much? The Cost of Alleviating Fuel Poverty*. Report to EAGA Partnership Charitable Trust, Bristol: EAGA.

Projecting Older People Population Information System (POPPI). 2010. *Population by Age* [Online]. Available at: www.poppi.org.uk/index.php?page No=314&areaID=8640&loc=8640 [accessed: 21 October 2010].

Rees Jones, I., Hyde, M., Victor, C.R., Wiggins, R., Gilleard, C. and Higgs, P. 2008. *Ageing in a Consumer Society: From Passive to Active Consumption in Britain*. Bristol: Policy Press.

Rowles, G.D. 1983. Geographical dimensions of social support in rural Appalachia, in *Ageing and Milieu: Environmental Perspectives on Growing Old*, edited by G.D. Rowles and R.J. Ohta. New York: New York Academic Press, 111-30.

Shah, H. and Marks, N. 2004. *A Wellbeing Manifesto for a Flourishing Society*. London: New Economics Foundation.

Wheelock, J. and Jones, K. 2002. Grandparents are the Next Best Thing: Informal Childcare for Working Parents in Urban Britain. *Journal of Social Policy*, 31(3), 441-64.

World Health Organization. 2007. *Global Age Friendly Cities: A Guide* [Online]. Available at: www.who.int/ageing/publications/Global_age_friendly_cities_Guide_English.pdf [accessed: 2 December 2007].

Chapter 6

The Role of Place Attachments in Wellbeing

Gordon Jack

Human experience is always rooted in place. (Entrikin 1989: 41)

If you ask any adult to talk about the factors that have shaped their identity, it is a good bet that they will talk not only about the most significant relationships and experiences in their life, but also about the places and wider geographical contexts within which these were set. As countless autobiographies, novels and films testify, the 'special places' of childhood, where children played and explored with their friends and had many of their formative experiences, as well as the wider geographical areas in which these experiences took place, tend to form a fundamental part of who people think they are (Chawla 1992). Meanwhile, in a seemingly parallel universe, the policy makers, organizations and professionals responsible for the wellbeing of some of the most disadvantaged and vulnerable children and young people in the UK (and elsewhere) paradoxically tend to almost completely ignore the role of people's attachments to places in the development of their personal identities. Although they recognize the central importance of attachment in human development, this is based almost exclusively on theories derived from developmental psychology which focus on the role of emotional attachments between children and their primary caregivers, and largely ignore attachments to place (e.g. Howe et al. 1999, Daniel et al. 1999, Aldgate et al. 2006). The inter-personal relationships deemed to be of central importance for children's wellbeing are treated as if they have developed in a vacuum.

In an attempt to address this deficit in current child welfare practice, insights primarily drawn from theory and research in the fields of human geography and environmental psychology are used to consider the meaning of place for individuals and the role that attachment to place plays in the development of personal identity (e.g. Proshansky et al. 1983, Corbishley 1995, Lalli 1992). Evidence about children's use of space and the implications for the development of their place attachments and associated wellbeing are then examined, and the chapter concludes by considering how to promote the place attachments of children and young people, especially those who spend significant periods of time in the care system.

The Meaning of Place and the Concept of Place Attachments

Place can be said to come into existence when people give meaning to a part of the larger, undifferentiated space in which they live. Whilst abstract knowledge about a place can be acquired relatively quickly, attachment to a place takes longer to develop, growing out of a large number of routine activities and everyday experiences, as well as more significant life events (Tuan 1977, Rowles 1983). Place exists at different scales, ranging from a particular part of a house or garden, to the streets, shops and other facilities and landmarks of the local neighbourhood or town, out to the wider countryside, region and nation. When people talk about where they 'feel at home' they might be referring to any or all of these levels (Tuan 1974, 1977).

Place attachment is a multi-faceted concept that characterizes the long-term affective bonds between individuals and significant places in their lives (Giuliani 2003, Low and Altman 1992). In an attempt to integrate the wide range of definitions of place attachment found within the literature, Scannell and Gifford (2010) have proposed an organizing framework that consists of three separate but overlapping dimensions – person, psychological process and place. In relation to the *person* dimension, they note that place attachments can operate at both individual and group levels, with the symbolic meanings of a particular place either being linked to individual experiences alone, or being shared among members of a group. The *psychological processes* involved in place attachment consist of a combination of emotional, cognitive and behavioural connections to particular places, exemplified by the feelings of sadness and longing which often follow major displacements. The final dimension is the nature of the *place* itself, which is normally sub-divided into physical ('rootedness') and social ('bondedness') components.

Place attachment is normally understood to be part of a person's overall identity (Proshansky et al. 1983, Corbishley 1995, Lalli 1992). Cognitions about places and the emotions associated with them become incorporated into the self, creating internalized objects that serve as sources of security at times of stress or isolation (Greenberg and Mitchell 1983, McCreanor et al. 2006). Places imbued with personal, social and cultural meaning therefore provide a framework within which important elements of personal identity are integrated (Cuba and Hummon 1993, Hay 1998). In the late modern/post-modern world, where tradition no longer defines the individual, people are constantly re-creating their roles and personal identities through reflexive processes of dynamic interaction between themselves and other members of the society in which they live (Giddens 1990, 1991). The risks and uncertainties created by the processes of modernization mean that attachment to place and the sense of belonging and security that it engenders can take on a particularly significant role in people's lives (Entrikin 1989).

The Origins of Place Attachments in Childhood

Place attachment depends on clusters of positive emotions and cognitions created by direct and repeated experiences of places in childhood, and the meanings associated with them. As a consequence, ideal settings may not result in place attachments if they are the site of particularly unhappy social interactions or life events, just as poor quality environments can produce place attachments if they are associated with positive experiences and memories (Morgan 2010). Furthermore, even when neither the physical environment nor the social context is ideal, resilient individuals are capable of transforming their experiences into positive cognitions (Proshansky et al. 1983).

As with human attachment, place attachment therefore grows out of regular cycles of person–environment interactions (Proshansky and Fabian 1987, Matthews 1992). Up to the age of four or five years, the home is likely to be the main setting for experiences of place, which have an unconscious, taken-for-granted quality at this stage of children's development. However, by the time they leave primary school, children will normally have developed some understanding of their geographical place in the wider world, supported by increasing knowledge of their local neighbourhood (Siegel and White 1975, Matthews 1992). Empirical evidence suggests that middle childhood is the time when the foundations of place attachments are laid down (Sobel 1990). Typically, this is followed in the teenage years by more conscious identification with the area in which a young person has grown up. Feelings of belonging tend to be strongest amongst young people who perceive that they have been fully included and accepted within their local community, or who have close connections through local ancestry (Gould and White 1968, Lynch 1977, Hay 1998, McCreanor et al. 2006).

Research with children and young people has repeatedly shown that place, identity and wellbeing are closely connected (Rowles 1980, 1983, Chawla 1992, Twigger-Ross and Uzzell 1996, Day and Milbjer 2007, Green and White 2007, Irwin et al. 2007). For example, a UK report that examined a wide range of evidence about the role of place in children's development concluded that their daily experience in the environment around them is a critical factor in their wellbeing (Sustainable Development Commission 2007: 6). The next sections therefore go on to examine children's access to their local environments, and how their use of space can affect the development of their place attachments.

Children's Use of Space and Place

Children's personal characteristics, family circumstances, and the wider environmental and cultural contexts in which they live all influence their use of space. Examination of these factors at individual, community and society levels, using an ecological framework, provides important insights into the way that different groups of children and young people use their surroundings, with varying implications for the development of their place attachments (Moore and Young 1978, Bronfenbrenner 1979, Jack 1997).

Individual/Family-level Influences

As already discussed, one of the main influences on children's use of space is their age, with many researchers noting the way that children typically progress from play close to home in the pre-school years, through increasing use of their local environment in middle childhood, to town-, or city-wide exploration during adolescence (Moore 1986, Matthews 1992, Chawla 1992). Work by Matthews (1987) in Coventry, for example, found a significant increase in the range of children's use of their local area between the ages of 8 and 9 years. This study, like many others (e.g. Jones et al. 2000), also noted that, with increasing age, parents tend to allow boys greater freedom to roam than girls.

The outdoor environments used by children depend on a number of factors besides their age, including their social class and the nature of their local surroundings (see, for example, Newson and Newson 1968, 1976). However, studies that have examined children's use of space indicate that their favourite places are homes and gardens, nearby streets and associated open spaces, parks, playgrounds and sports fields (Moore 1986, Gill and Sharps 2004, Burke 2005). Several studies also emphasize the particular significance of unsupervised 'secret places' that facilitate imaginative play (Lynch 1977, Moore 1986, Thomas and Thompson 2004).

Community/Neighbourhood-level Influences

The levels of safety and risk posed by the local environment, as assessed by both parents and children, also influence children's use of space (Garbarino et al. 1992, Thomas and Thompson 2004, Synovate 2010). For example, one study conducted in Belfast found that the mobility of seven to eight-year-olds was severely limited by parental fears of sectarian violence (Connolly and Neill 2001); whilst another study conducted in the same city with teenagers found that, although their range of exploration was much wider (and direct parental intervention was less apparent) the young people themselves imposed limits on their daily movements, based on their own perceptions of safety and risk (Leonard 2007). Other studies have revealed the way that assessments of the dangers posed by traffic and crime in different locations mean that children living in inner-city or urban locations typically enjoy less independent mobility than their peers living in suburban or rural areas (Deakin 2006, Morrissey and Smyth 2002). For example, a study of 13 and 14-year-olds living in three different locations in the English Midlands found that those living in urban areas were less likely to travel unaccompanied than their peers in suburban and rural areas, where lone travel, even after dark, was quite common (Jones et al. 2000). Fears about crime and traffic also have negative impacts on young people's perceptions of their local area, affecting not only their views about its safety, but also about the level of friendliness and helpfulness of local people (Morrow 2002, Mullan 2003).

Children's journeys to and from school are also strongly influenced by parental perceptions about risks, and a number of studies in the UK have revealed the declining frequency with which children are allowed to travel to school unaccompanied by an adult (Hillman, et al. 1990, O'Brien 2000). This is important because research has highlighted the significant role that children's independent journeys to school play in the development of their place attachments. Ross (2007) for example, found that children's unaccompanied journeys to school, together with the freedom to use local areas in a relatively unstructured way and to visit favourite places and people independently, facilitated a wide range of social and environmental interactions with people and places, contributing to children's personal and community identities. These independent experiences of place gave children a visible presence in their local area, facilitating the development of 'weak ties' with other members of their community which enhanced their feelings of belonging and security (Granovetter 1973, Ross 2007).

Society/Cultural-level Influences

Children's use of space is also strongly influenced by the rules and meanings that exist within particular cultures (Rappaport 1978). Early empirical evidence suggested that an understanding of children's behaviour depended more on knowledge about the cultural rules governing different settings than children's individual characteristics (Barker and Wright 1951). More recent research supports this conclusion. For example, when 10 and 11-year-olds living in four locations in England were asked about their use of outdoor space, they repeatedly referred to the influence of the social codes and expectations that existed in their local communities (Thomas and Thompson 2004). There is also an association between children's behaviour and the cultural artefacts available to them. For example, the increasing access that most children in the UK have to different forms of home entertainment, including TVs, video games and computers, in combination with parental fears about traffic, crime and 'stranger danger', has resulted in more of their time being spent indoors (Sustainable Development Commission 2007).

The cultural norms that influence children's behaviour in community settings are generally established, maintained and transmitted by adults, who often take steps to manage children's use of space through a combination of surveillance and regulation, including the employment of staff to patrol neighbourhoods and shopping centres, and the erection of signs prohibiting activities such as cycling and playing ball games (Holland et al. 2007). Restrictions like these are often perceived by children and young people to be unfair, especially as they usually lack effective means to challenge them (Morrow 2002). Most children also report having been told off by neighbours for playing outside, with a third of seven to 11-year-olds in one UK survey, for example, saying that this had stopped them playing outdoors (Stockdale et al. 2003).

Unfortunately, as the next section goes on to consider, the range of children's outdoor activities that adults in the UK perceive to be too risky or socially unacceptable is becoming ever more restrictive, even though these judgements often bear little relation to the likelihood of actual harm or disturbance (Play England 2008).

The Shrinking World of Childhood

It can be argued that the decline in children's independent journeys to and from school witnessed in most parts of the UK over recent decades is indicative of a wider risk-averse and intolerant society. Several researchers and commentators have drawn attention to the 'protective discourse' that has developed over this period, with growing concerns about children's safety progressively reducing their freedom to use the local environment (e.g. Scott et al. 1998, Morrow 2002, Gill 2008). Nearly half of adults consulted in one survey, for example, expressed the view that children under 14 years of age should not be allowed to go outside with their friends without adult supervision (Children's Society 2007). An opinion poll conducted with children and young people about risk and play also revealed that many 'normal' childhood activities, such as climbing trees, playing in the local park or streets and riding a bike to a friend's house, were not permitted by significant proportions of parents unless children were supervised by an adult (Play England 2008).

A number of UK studies of children's experiences and adults' attitudes have highlighted the way that interactions between factors within this protective discourse contribute to what is referred to as the 'shrinking world of childhood' (e.g. Gill 2008). Although the growth in home entertainment noted in the last section has undoubtedly made a significant contribution to the decline in children's use of outdoor space, this is clearly not the whole story as surveys also reveal that most children would like to spend more time outdoors (e.g. Thomas and Thompson 2004). Restrictions on children's freedom of use of the local environment are also dependent on factors in the environment itself, including a decline in public open space available to children (Thomas and Thompson 2004, Westbury 2007).

Although restrictions on children's freedom are generally imposed in the name of protecting them from harm, there is growing evidence about the damaging effects that they may actually be having on children's wellbeing, evident, for instance, in rising levels of childhood obesity and mental health problems (Gaster 1991, Armstrong 1993, Mental Health Foundation 1999, Collishaw et al. 2004, Irwin et al. 2007). Support for this conclusion comes from studies examining the effects of increased access to the outdoors, which has been shown to reduce feelings of anger and anxiety (Pretty et al. 2005), and improve the concentration and self-discipline of children with conduct disorders (Taylor et al. 2001).

Less obvious, but arguably just as harmful, is the effect that the increasingly restricted use of the local environment is having on the development of children's place attachments. Particularly during middle childhood, when the foundations of place attachments are laid down (Sobel 1990), the shrinking world of childhood means that children are likely to have much less independent experience of place, thereby limiting the development of the place attachments which are known to be important for the development of their identity and sense of security and belonging. In this context, the following section considers how children's place attachments can be promoted, reviewing relevant strands of government policy, and identifying some of the ways that the additional problems faced by children in the care system might be addressed.

Promoting the Place Attachments of Children and Young People

Social Policy for Children

Three broad strands of social policy relevant to the promotion of children's wellbeing can be identified (Gill 2008). Within the *laissez-faire* strand, the state's role in relation to the majority of children and young people is confined to funding a range of universal services, such as schools and health care (Department for Children, Schools and Families 2007). The second, *service-oriented* strand of policy consists of specific programmes designed to improve particular aspects of child wellbeing. Recent examples relevant to children's use of space include the Healthy Schools and the Travelling to School programmes, both of which promoted the benefits of walking and cycling to school. However, neither of these two strands of social policy is likely to increase children's independent use of their local environment, which is the key to developing their sense of security and belonging in their local communities (Engwright 2008). Only the third, *space-oriented* strand of policy specifically aims to do this, through the development of what are often referred to as 'child-friendly communities' (e.g., Arnold and Cloke 1998, Gill 2008). These are communities that attempt to strike a balance between the protection and freedom of children, accepting a shared sense of responsibility for their wellbeing, viewing children as capable of shaping their own lives, and providing local environments that are attractive and safe, with plenty of opportunities for play and informal social interaction (Verwer 1980, Moore 1986).

In England, evidence of the space-oriented strand of policy is most apparent in the Children's Plan (Department for Children, Schools and Families 2007) and the accompanying Play Strategy, which stated that 'strong, vibrant communities should offer a variety of places for children to play, places in which children have a stake, and that they can help shape through their active involvement in design and decision making. All children and young people should be able to find places, near their homes, where they can play freely and meet their friends' (Department for Children, Schools and Families/Department for Culture, Media and Sports 2008: 5). Concrete expression of this strand of policy can be found in a number of generally small-scale government-supported projects, such as the

Neighbourhood Play Toolkit, developed by the Children's Play Council, which provides a range of resources for adults and children to use together, designed to improve children's access to safe neighbourhood play spaces (Kapasi 2006). Grants from the Department of Transport have also supported the creation of a number of partnerships between local authorities and groups of residents who want to develop 'home zones' – small residential areas in which pedestrians and cyclists are given priority over cars, encouraging children's play and other forms of social activity and interaction between local people (Biddulph 2001). Areas where successful traffic calming or pedestrian-oriented measures like home zones have been introduced demonstrate an increase in informal social activities and children's play (British Medical Association 1997, Pharaoh and Russell 1989). However, it has to be recognized that, overall, this aspect of social policy is not well established or funded in the UK, and is likely to be vulnerable in the context of the major cuts in public spending planned to take effect from 2010 onwards.

Children and Young People in the Care System

At any time there are approximately 60,000 children and young people being looked after away from home by local authorities in England (Department for Children, Schools and Families 2010). For obvious reasons, the place attachments of these children and young people, separated from their families of origin and familiar surroundings, are particularly vulnerable. Children who experience repeated moves have been found to display a sense of rootlessness and a fragmented identity (Coles 1970), and adults who experience enforced separations from their familiar surroundings have a diminished sense of self, with increased levels of anxiety and vulnerability to depression (Twigger-Ross and Uzzell 1996, Korpela 2002, Vandemark 2007).

The long-term effects of major displacements in people's lives depend upon a number of factors, including the strength their original place attachments, the degree to which the significance of their loss is recognized, the extent of the changes involved, and the availability of support and opportunities for positive action after the dislocation (Young and Wilmott 1957, Fried 1963, Edelstein 1988). Social workers and other professionals working with children and young people in the care system often undertake 'life-story work' to help them understand the dislocations that they have experienced, often including several changes of home and carers. However, the literature currently available to guide practitioners in this task (e.g. Fahlberg 1991, Ryan and Walker 2007) makes no specific mention of children's place attachments. Based on the evidence reviewed in this chapter so far, it is clear that this is a serious omission which needs to be addressed.

One of the ways in which greater recognition of the role of place attachments in children's wellbeing can be translated into child welfare practice with children and young people, whether separated from their families of origin or not, is to talk to them more about the places that are important or special to them. Helping children and young people to recall memories and develop stories that express the meaning of different places to them, and the role that they play in their sense of who they are, could make a significant contribution to their wellbeing. However, as already discussed, increasing fears about the safety of the local environment and the attraction of different forms of home entertainment mean that children's direct experience of their surroundings is declining. In this context, simply talking to children, especially those who spend significant periods of time in care, about their experiences of place is unlikely to be sufficient on its own. What is also required is a conscious effort to expand the range of day-to-day experiences of place available to them.

For younger children who experience prolonged periods in care, this combination of approaches might entail ensuring that they have repeated opportunities to experience their local environment in the company of adults or older children, by regularly walking to visit relatives, friends, shops, libraries, playgrounds and parks. During these outings, children should be allowed opportunities to play, explore and daydream, as well as being introduced to local people along the way whenever this is appropriate. Those accompanying them should also point out landmarks, and talk about the places that are significant to the child, subsequently incorporating them into family stories and photograph albums or life-story books (Bohanek et al. 2006). Older children in care for significant periods of time need increasingly independent opportunities to use and explore their local surroundings, including journeys to and from school, socializing with friends, and using local facilities. Caregivers should work gradually towards giving the young people they are looking after full independence, through a step-by-step process involving diminishing levels of adult supervision. Once again, the significance of place within these experiences needs to be recognized, discussed and incorporated into the young person's developing sense of self, helping them to make meaningful links between their past, present and future. For children of all ages looked after away from home for prolonged periods of time, regular experience of being out and about in their local surroundings will also help to give them the sort of 'visible presence' that is key to the development of a sense of belonging (Ross 2007). These experiences also provide the basis for potentially rich and rewarding inter-personal interactions, which will help to build secure emotional attachments in the home, thereby promoting children's resilience (Newman 2004).

Finally, the importance of children's increasingly independent use of their surroundings as they grow older also draws attention to the specific characteristics of different local areas that make them more or less safe and attractive for children and young people to use. Effective promotion of the place attachments of looked after children therefore also requires that those who are responsible for their care take an active part in initiatives and campaigns that aim to develop aspects of more child-friendly communities. This means that social workers and others working in children's services need to develop strategies of their own, as well as actively supporting and participating in initiatives set up by others, which:

- view children and young people as capable of shaping their own lives.
- aim to improve the safety of local areas for children's play, exploration and social interaction.
- develop the understanding of local people about children's needs, enabling them to strike a balance between their protection and freedom; and
- help communities to accept shared responsibility for children's wellbeing.
- Conclusion.

Despite significant increases in welfare spending by the Labour Government during the period from 1997 to 2010, the wellbeing of children and young people in the UK, using measures of their health, education and social relationships, remained stubbornly low in comparison to other developed countries around the world during this period (Bradshaw et al.2006, UNICEF 2007). At the time of writing, in the aftermath of the global financial crisis and a period of economic recession, the coalition government formed in the UK in 2010 plans to tackle rising levels of national debt by making radical cuts to public spending which are likely to further undermine child wellbeing.

These circumstances present policy makers, organizations and professionals charged with responsibility for the welfare of children with significant challenges. The solution to these challenges has to lie primarily in longer-term economic recovery and renewed government commitments to reducing inequalities in income and other forms of disadvantage. However, in the shorter term, the evidence presented in this chapter suggests that greater recognition of the role that place attachments play in the development of personal identity, including increased feelings of security and belonging, has the potential to bring about significant improvements in child (and adult) wellbeing.

References

Aldgate, J., Jones, D., Rose, W. and Jeffery, C. 2006. *The Developing World of the Child*. London: Jessica Kingsley.

Armstrong, N. 1993. Independent mobility and children's physical development, in *Children, Transport and the Quality of Life*, edited by M. Hillman. London: Policy Studies Institute, 35-43.

Arnold, E. and Cloke, C. 1998. "Society keeps abuse hidden—the biggest cause of all": The case for child-friendly communities. *Child Abuse Review*, 7, 302-14.

Barker, R.G. and Wright, H.F. 1951. *Midwest and its Children*. New York: Row, Peterson & Company.

Biddulph, M. 2001. *Home Zones: A Planning and Design Handbook*. Bristol: The Policy Press.

Bohanek, J.G., Marin, K.A., Fivish, R. and Duke, M.P. 2006. Family narrative interaction and children's sense of self. *Family Process*, 45(1), 39-54.

Bradshaw, J., Hoelscher, P. and Richardson, D. 2006. An index of child wellbeing in the European Union. *Social Indicators Research*, 80(1), 133-77.

British Medical Association. 1997. *Road Transport and Health*. London: BMA.

Bronfenbrenner, U. 1979. *The Ecology of Human Development*. Cambridge, MA: Harvard University Press.

Burke, C. 2005. "Play in focus": Children researching their own spaces and places for play. *Children, Youth and Environments*, 15(1), 27-53.

Chawla, L. 1992. Childhood place attachments, in *Place Attachment*, edited by I. Altman and S.M. Low. New York: Plenum Press, 63-86.

Children's Society. 2007. *Reflections on Childhood: Friendship*. London: Children's Society.

Coles, R. 1970. *Uprooted Children*. Pittsburgh: Pittsburgh University Press.

Collishaw, S., Maughan, B., Goodman, R. and Pickles, A. 2004. Time trends in adolescent mental health. *Journal of Child Psychology and Psychiatry*, 45, 1350-62.

Connolly, P. and Neill, J. 2001. Constructions of locality and gender and their impact on the educational aspirations of working-class children. *International Studies in the Sociology of Education*, 11(2), 107-29.

Corbishley, P. 1995. A parish listens to its children. *Children's Environments*, 12, 414-26.

Cuba, L. and Hummon, D.M. 1993. A place to call home: Identification with dwelling, community and region. *The Sociology Quarterly*, 34(1), 111-31.

Daniel, B., Wassell, S. and Gilligan, R. 1999. *Child Development for Child Care and Protection Workers*. London: Jessica Kingsley.

Day, C. and Midbjer, A. 2007. *Environments for Children: Passive Lessons from the Everyday Environment*. Oxford: Architectural Press.

Deakin, J. 2006. Dangerous people, dangerous places: The nature and location of young people's victimisation and fear. *Children and Society*, 20, 376-90.

Department for Children, Schools and Families. 2007. *The Children's Plan: Building Brighter Futures*. London: DCSF.

Department for Children, Schools and Families. 2010. *Children Looked After in England Year Ending 31 March 2009*. London: DCSF.

Department for Children, Schools and Families/Department for Culture, Media and Sport. 2008. *Fair Play: A Consultation on the Play Strategy*. London: DCSF.

Edelstein, M.R. 1988. *Contaminated Communities: The Social and Psychological Impacts of Residential Toxic Exposure*. Boulder, CO: Westview Press.

Engwright, D. 2008. *Is the Walking School Bus Stalled in an Evolutionary Cul-de-Sac?* [Online]. Available at: www.lesstraffic.com/Articles/Traffic/wbstalled. htm [accessed: 16 April 2008].

Entrikin, J.N. 1989. Place, region and modernity, in *The Power of Place: Bringing Together Geographical and Sociological Imaginations*, edited by J.A. Agnew and J.S. Duncan. London: Unwin Hyman, 30-43.

Fahlberg, V. 1991. *A Child's Journey through Placement*. London: BAAF.

Fried, M. 1963. Grieving for a lost time, in *The Urban Condition*, edited by L.J. Duhl. New York: Basic Books, 151-71.

Garbarino, J., Dubrow, N., Kostelny, K. and Pardo, C. 1992. *Children in Danger.* San Francisco, CA: Jossey-Bass.

Gaster, S. 1991. Urban children's access to their neighborhood: Changes over three generations. *Environment and Behavior*, 23, 70-85.

Giddens, A. 1990. *The Consequences of Modernity.* Cambridge: Polity Press.

Giddens, A. 1991. *Modernity and Self-Identity.* Cambridge: Polity Press.

Gill, O. and Sharps, A. 2004. *Children and Community in Central Weston-Super-Mare.* Bristol: Barnardo's.

Gill, T. 2008. Space-oriented children's policy: Creating child-friendly communities to improve children's wellbeing. *Children and Society*, 22, 136-42.

Giuliani, M.V. 2003. Theory of attachment and place attachment, in *Psychological Theories for Environmental Issues*, edited by M. Bonnes, T. Lee and M. Bonaluto. Aldershot: Ashgate, 137-70.

Gould, P. and White, R. 1968. The mental maps of British school leavers. *Regional Studies*, 2, 161-82.

Granovetter, M.S. 1973. The strength of weak ties. *American Journal of Sociology*, 78(6), 1360-80.

Green, A.E. and White, R.J. 2007. *Attachment to Place: Social Networks, Mobility and Prospects of Young People.* York: Joseph Rowntree Foundation.

Greenberg, J.R. and Mitchell, S.A. 1983. *Object Relations in Psychoanalytic Theory.* Cambridge, MA: Harvard University Press.

Hay, R. 1998. Sense of place in developmental context. *Journal of Environmental Psychology*, 18, 5-29.

Hillman, M., Adams, J. and Whitelegg, J. 1990. *One False Move: A Study of Children's Independent Mobility.* London: Policy Studies Institute.

Holland, C., Clark, A., Katz, J. and Peace, S. 2007. *Social Interactions in Urban Public Places.* Bristol: The Policy Press.

Howe, D., Brandon, M., Hinings, D. and Schofield, G. 1999. *Attachment Theory, Child Maltreatment, and Family Support: A Practice and Assessment Model.* Basingstoke: Macmillan.

Irwin, L.G., Johnson, J.L., Henderson, A., Dahinten, V.S. and Hertzman, C. 2007. Examining how contexts shape young children's perspectives of health. *Child: Care, Health and Development*, 23(4), 353-59.

Jack, G. 1997. An ecological approach to social work with children and families. *Child and Family Social Work*, 2(2), 109-20.

Jones, L., Davis, A. and Eyers, T. 2000. Young people, transport and risk: Comparing access and independent mobility in urban, suburban and rural environments. *Health Education Journal*, 59, 315-28.

Kapasi, H. 2006. *Neighbourhood Play and Community Action.* York: Joseph Rowntree Foundation.

Korpela, K.M. 2002. Adolescents' favourite places and environmental self-regulation. *Journal of Environmental Psychology*, 12, 249-58.

Lalli, M. 1992. Urban-related identity: Theory, measurement and empirical findings. *Journal of Environmental Psychology*, 12(4), 285-303.

Leonard, M. 2007. Trapped in space? Children's accounts of risky environments. *Children and Society*, 21, 432-45.

Low, S.M. and Altman, I. 1992. Place attachment: A conceptual inquiry, in *Place Attachment*, edited by I. Altman and S.M. Low. New York: Plenum Press, 1-12.

Lynch, K. 1977. *Growing Up in Cities*. Cambridge, MA: MIT Press.

Matthews, M.H. 1987. Gender, home range and environmental cognition. *Transactions of the Institute of British Geographers, New Series*, 12, 234-39.

Matthews, M.H. 1992. *Making Sense of Place: Children's Understanding of Large-Scale Environments*. Hemel Hempstead: Harvester Wheatsheaf.

McCreanor, T., Penney, L., Jensen, V., Witten, K., Kearns, R. and Barnes, H.M. 2006. "This is like my comfort zone": Senses of place and belonging within Oruamo/Beachhaven, New Zealand. *New Zealand Geographer*, 62(3), 196-207.

Mental Health Foundation. 1999. *Brighter Futures: Promoting Children and Young People's Mental Health*. London: MHF.

Moore, R. 1986. *Childhoods Domain*. London: Croom Helm.

Moore, R.C. and Young, D. 1978. Childhood outdoors: Toward a social ecology of the landscape, in *Children and the Environment*, edited by I. Altman and J. F. Wohlwill. New York: Plenum Press, 83-130.

Morgan, P. 2010. Towards a developmental theory of place attachment. *Journal of Environmental Psychology*, 30(1), 11-22.

Morrissey, M. and Smyth, M. 2002. *Northern Ireland after the Good Friday Agreement: Victors, Grievance and Blame*. London: Pluto Press.

Morrow, V. 2002. Children's rights to public space: Environment and curfews, in *The New Handbook of Children's Rights*, edited by B. Franklin. London: Routledge, 168-81.

Mullan, E. 2003. Do you think that your local area is a good place for young people to grow up? The effects of traffic and car parking on young people's views. *Health and Place*, 9, 351-60.

Newman, T. 2004. *What Works in Building Resilience?* Barkingside: Barnardo's.

Newson, J. and Newson, E. 1968. *Four Years Old in an Urban Community*. London: Allen & Unwin.

Newson, J. and Newson, E. 1976. *Seven Years Old in the Home Environment*. London: Allen & Unwin.

O'Brien, M. 2000. *Childhood, Urban Space and Citizenship: Child-Sensitive Urban Regeneration*, ESRC Children 5-16 Research Briefing No. 16. Swindon: ESRC.

Pharaoh, T. and Russell, J. 1989. *Traffic Calming: Policy and Evaluations in Three European Countries*, Occasional Paper 2/1989. London: Department of Planning, Housing and Development, South Bank Polytechnic.

Play England. 2008. *Playday 2008 Opinion Poll Summary* [Online]. Available at: www.playengland.org.uk [accessed: 15 July 2008].

Pretty, J., Peacocok, J., Sellens, M. and Griffin, M. 2005. The mental and physical health outcomes of green exercise. *Journal of Environmental Health Research*, 15(5), 319-37.

Proshansky, H.M. and Fabian, A.K. 1987. The development of place identity in the child, in *The Built Environment and Child Development*, edited by C.S. Weinstein and T.G. David. New York: Plenum Press, 21-40.

Proshansky, H.M., Fabian, A.K. and Kaminoff, R. 1983. Place identity: Physical world socialization of the self. *Journal of Environmental Psychology*, 3, 57-83.

Rappaport, A. 1978. On the environment as an enculturating system, in *Priorities for Environmental Design Research*, edited by S. Weidmann and J. A. Anderson. Washington: EDRA, 273-86.

Ross, N.J. 2007. "My journey to school": Foregrounding the meaning of school journeys and children's engagements and interactions in their everyday localities. *Children's Geographies*, 5(4), 373-91.

Rowles, G.D. 1980. Growing old "inside": Aging and attachment to place in an Appalachian community, in *Transitions of Aging*, edited by N. Datan and N. Lohmann. New York: Academic Press, 153-70.

Rowles, G.D. 1983. Place and personal identity in old age: Observations from Appalachia. *Journal of Environmental Psychology*, 3, 299-313.

Ryan, T. and Walker, R. 2007. *Life Story Work*, 3rd Edition. London: BAAF.

Scannell, L. and Gifford, R. 2010. Defining place attachment: A tripartite organizing framework. *Journal of Environmental Psychology*, 30(1), 1-10.

Scott, S., Jackson, S. and Backett-Milburn, K. 1998. Swings and roundabouts: Risk, anxiety and the everyday worlds of children. *Sociology*, 32(4), 689-705.

Siegel, A.W. and White, S.H. 1975. The development of spatial representations of large-scale environments, in *Advances in Child Development and Behavior*, edited by H.W. Reese. New York: Academic Press, 241-62.

Sobel, A. 1990. A place in the world: Adults' memories of childhood's special places. *Children's Environments Quarterly*, 6(4), 25-31.

Stockdale, D., Kafz, A. and Brook, L. 2003. *You Can't Keep Me In!* London: The Children's Society/Young Voice.

Sustainable Development Commission. 2007. *Every Child's Future Matters*. London: SDC.

Synovate (UK) Ltd. 2010. *Staying Safe Survey 2009: Young People and Parent's Attitudes Around Accidents, Bullying and Safety.* London: Department for Children, Families and Schools.

Taylor, A., Kuo, F. and Sullivan, W. 2001. Coping with ADD: The surprising connection to green play settings. *Environment and Behavior*, 33, 54-77.

Thomas, G. and Thompson, G. 2004. *A Child's Place: Why Environment Matters to Children*. London: Green Alliance/Demos.

Tuan, Y.F. 1974. *Topophilia: A Study of Environmental Perception, Attitudes and Values*. Englewood Cliffs, NJ: Prentice Hall.

Tuan, Y.F. 1977. *Space and Place: The Perspective of Experience*. Minneapolis, MN: University of Minneapolis Press.

Twigger-Ross, C.I. and Uzzell, D.L. 1996. Place and identity processes. *Journal of Environmental Psychology*, 16, 205-20.

UNICEF. 2007. *Child Poverty in Perspective: An Overview of Child Wellbeing in Rich Countries*. Florence: UNICEF Innocenti Research Centre.

Vandemark, L.M. 2007. Promoting the sense of self, place, and belonging in displaced persons: The example of homelessness. *Archives of Psychiatric Nursing*, 21(5), 241-8.

Verwer, D. 1980. Planning residential environments according to their real use by children and adults. *Ekistics*, 281, 109-13.

Westbury, P. 2007. *A Sense of Place: What Residents Think of their New Homes*. London, CABE.

Young, M. and Wilmott, P. 1957. *Family and Kinship in East London*. New York: Penguin.

Chapter 7

Am I an Eco-Warrior Now? Place, Wellbeing and the Pedagogies of Connection

Andrea Wheeler

Changing notions of place, informed by concurrent philosophical perspectives, have always played an important role in architectural theory. The concepts of design for the purposes of wellbeing and the relationship of design with sustainability are, however, more contemporary issues within the influences on architectural thinking. What links such notions are ideas of place and identity. Calls for less materialistic approaches to how we are in the world are informed by debates over what constitutes present and future wellbeing and in particular by those concerned with climate change and the nature of happiness. In turn, these debates interlink with questions of the relationship between design and individual action in the form of changes in lifestyle for current and future, sustainable wellbeing. Government initiatives and media campaigns distribute information about the need to reduce our levels of consumption, and policy makers argue over the best way to motivate people (Collier et al. 2010), but our culture imposes a pressure to continue to consume. In this context, it is not surprising that many young people feel confused about their personal responsibility in adapting to climate change.

Property and material wealth provide identity, social status and a feeling of belonging; satisfying the apparent need for new consumer goods gives pleasure, fleeting as it may be. Manufacturers are keen to exploit such needs and feelings through the continuous promotion of novel consumer products directed at young people and designers are all too willing to perpetuate the age old vagaries of fashion. In deprived communities such material aspirations have very real consequences for the health, wellbeing and life chances of families and young people. And yet those people rarely change their behaviour only on the basis of rational calls to do so; knowledge and beliefs do not simply and straightforwardly lead to actions that are more or less environmentally friendly, ethical or appropriate. The need for education for pro-environmental and sustainable behaviour is becoming ever more pressing. The school and its design affords an interesting location for such debates. This chapter explores the rhetoric, experiences and issues in developing designs for sustainable wellbeing in the school setting and refers to research carried out with children in schools in the Midlands and North of England.

In 2004, the UK government created a unique educational opportunity with its new school building programme. Its aims were to transform learning and embed sustainability into the life experience of every child. Participation was at the forefront of the initiative and was intended to encourage a sense of community, ownership and belonging. Participation does not however necessarily mean equity or inclusion, nor does it guarantee that any greater care will be taken of the immediate environment of the school or, ultimately, the global environment. Participation will neither, as a matter-of-course, lead to a greater connection with the school environment with others, nor a greater sense of wellbeing. Why? Because at the heart of the problem are deep rooted social and cultural norms.

The Building Schools for the Future (BSF) programme outlined plans to rebuild or refurbish every secondary school in England over a 15-20 year period, targeting local authorities with the most deprived schools first. The environmental ambitions of the programme were evident from the outset:

> Sustainable development will not be just a subject in the classroom: it will be in its bricks and mortar and in the way the school uses and even generates its own power. Our students won't just be told about sustainable development: they will see and work within a school that is a living, learning place in which to explore what a sustainable lifestyle means. (Blair 2004)

However, in 2010 the new UK coalition government cancelled the BSF programme and only schools that had already signed contracts would go ahead with construction. At the point when the programme was cancelled, 185 schools had received BSF funding. But even without the school building programme providing the platform for social change, the UK government still requires buildings to contribute to meeting carbon emission reduction targets and this aim will only be met if the occupants believe in the need to reduce energy consumption. Integrating environmental awareness and behaviours with design of the built environment is still a highly significant issue for reducing carbon emissions and schools provide a unique set of circumstances for exploring these relationships.

Zero-Carbon Schools?

Promoting a sustainable notion of wellbeing has been supported in intention, alongside the BSF programme, through the UK Zero Carbon Schools Task Force. This was established in 2008 by the Secretary of State for Children, Schools and Families and operated until 2009 with a remit to advise on what needs to be done in order to reach the stated goal that all new school buildings will be carbon neutral by 2016. As part of that brief, the Task Force identified problems in the energy performance of new schools and teased out the central challenges in developing 'zero carbon schools'. The important place of schools in the move towards zero carbon and more sustainable communities and the role of children's agency in achieving this transition were seen as critical:

> Schools are crucial in achieving lower energy ambitions, not least because of so many students' enthusiasm for helping to protect the future of the planet. And it is not just the students; it is their families, their homes and their communities that surround the schools. (Department for Children, Schools and Families 2010a: 2)

Furthermore, the Task Force elaborated five steps towards making this happen, described as 'The Carbon Hierarchy': engagement with school communities; reducing demand (assisted by engagement leading to changes in behaviour); driving out waste by better design (which will need more knowledge and skills in the design and construction industries); decarbonizing school energy supplies; and neutralizing any residual emissions (DCSF 2010a). The Task Force argued that low and zero carbon buildings will only be achievable if action is taken across a range of fronts including technical, financial and social and that this social dimension will include critical engagement of children and leadership by teachers supported by local authorities. The Task Force reported during a period of intensive new building; as already indicated, what has followed is a climate of cuts in public funding (including the BSF programme) meaning that behaviour-driven factors may become even more significant in reducing carbon emissions. Indeed, the Task Force argued that much can be done without capital investment in new schools and that retrofit may have a far greater impact than a single focus on new build:

> With the active engagement of students and staff and an understanding of how and where energy is being used, all schools can reduce their energy use. But this requires leadership from headteachers, governors and other decision makers, with appropriate support and guidance from local authorities. (Department for Children, Schools and Families 2010a: 4)

The report also recommended that the agency with responsibility for delivering the BSF programme, Partnership for Schools, should develop both a post-occupancy evaluation (POE) process for all schools within BSF and a methodology for an in-depth energy study to be applied annually to a sample of schools [Recommendation 25] (DCSF 2010a: 8). Other recommendations include: the gathering and publication of performance data in order to monitor progress [Recommendation 26]; a targeted programme of energy-reducing refurbishment work (linked to behavioural change) in order to cut emissions in existing schools [Recommendation 27]; education and engagement initiatives for staff, students and communities [Recommendations 3, 4, 5]. All of these recommendations promote the need for a continuous educational cycle of feedback, monitoring and action to achieve a reduction in carbon emissions – in effect a learning process towards more sustainable lifestyles.

With the new wave of school building, and an on-going need to retrofit old buildings, children will grow up within planned environments that pay significant attention to the idea of reducing energy consumption. However, whilst architects strive to produce sustainable and low carbon designs, the reality emerging is somewhat different as the new schools are consuming substantially higher levels of energy than anticipated. The new schools have proved disappointing for several technical reasons: the building fabric performance is not always as good in practice as in theory; the building systems and controls are too complicated; the demand-responsiveness to patterns of use is poor, and such unmanageable complications lead to avoidable waste (Bordass 2009).

However conflicting social factors, most notably the behaviour of the occupants, alongside the trend for educational technology, also cause increasingly intensive use of energy. For example, interactive whiteboards undermine daylight strategies, ICT policies, with the aim of having one computer per student, increases electricity consumption and extended hours of use of the building for community activities also increases electricity consumption. Dysfunctional procurement methods also make it difficult to pay any attention to the detail of a building's performance or to adopt strategies which can integrate consideration of both technological solutions and human behaviours (Bordass 2009).

Whilst many of the more hidden energy efficient design strategies go unnoticed by children and adults alike, children are quick to point out many of the more obviously wasteful energy behaviours happening in otherwise energy efficient schools (and many of these are more often than not directed by adults' rules). In a recent series of post-occupancy workshops with children in new schools such problems are evident, as shown in dialogue from pupils aged 13 years old:

> Pupil 1: I think we should stop lighting the school in the day as the sun lights it up a lot and we're wasting electricity.

> Pupil 2: All the computers are always on, they are never switched off by the power. They are always on standby. [...] it's just that the monitor is off. You just logoff and you don't shut it down.

> Pupil 3: They are telling us to be energy efficient but... They stand there in science and say you need to save energy and then I say well turn your lights off... they are always banging on about it. They are always telling us to save energy but why not them.

Children are aware of the issue of energy efficiency and the problem of lifestyle change and in many instances their familiarity with their environment means they can devise good solutions. The combination of cancelling the building programme and the disappointing results from new schools to date, both in terms of qualitative and qualitative data, indicates that achieving government targets for carbon emission reduction will only be met if the behaviours of the occupants of buildings can be changed. The early rhetoric of the BSF programme, highlighting the need to integrate environmental awareness and behaviours, is thus still relevant in reducing carbon emissions. Although new schools may not be essential in a difficult financial climate, a new way of thinking about the school environment and the relationship between sustainability and educational aims remains so. Behavioural change is an issue for educational theory, and yet such questions are rarely considered within the broader field of energy behaviours.

In response to some of these issues, Bunn (2009) argues for the 'humane design' of sustainable schools, an approach he defines as ergonomic and democratic: that is, a design solution that meets users' needs rather than the designers' beliefs or ideas of what teachers ought to have, regardless of whether they really want or need it. He argues, for example 'hand-held remotes have been given to school-appointed eco-warriors to control lights. Pupil power can be as powerful as BEMS (Building Environmental Management Systems) when it comes to truly intelligent lighting control' (Bunn 2009: 41). Similarly, Bordass (2009) does not argue for technological solutions or more intelligent design in order to improve building performance, but rather that engaging people in the problem could halve the overall energy demand.

Hence, the need for integrated approaches to addressing both technical performance and occupant behaviour is becoming increasingly acknowledged even within the technical field. As Stevenson argues 'without knowledge of both technical performance and occupant behaviour, it will not be possible to optimize design or to predict actual performance with [any] reliability' (2010: 436). In new housing, 'rebound' effects are now widespread; people are being provided with energy efficient, well-insulated homes and, as a result, come to demand higher levels of comfort and thus, perversely, use more energy than in the old houses (Stevenson 2010). However, whilst a focus on technological features in sustainable schools will provide answers to performance questions, engaging people (let alone children) in sustainable behaviours is far from simple (Newman et al. 2009, Chernely 2010).

Bordass, Bunn and Stevenson sit within a growing field of studies related to post-occupancy assessments of buildings. This new focus explores the contribution made by lifestyle factors, social norms and culture, to the energy performance of buildings (Gill et al. 2010, Gram-Hansen 2010, Stevenson 2008). In schools themselves, attention is given to how culture and ethos motivate the behaviours of young people. Addressing a sustainable energy agenda might sit well within this work. But although encouraging sustainable lifestyles is a contemporary policy issue, in schools, education for sustainability tends to have a low priority and teachers lack incentives to act in accordance with this educational message, which children note sometimes with annoyance. Different ideas towards energy efficiency will emerge in the younger generation, whether as a result of an architectural environment or other factors, and it is important that schools act to reinforce emerging lifestyles, and be more critical of adults' habits and 'old ways'. Indeed, Sanoff (1992) argues that school culture appears to have the strongest influence on attitudes to change, and school managers and heads in failing schools are increasingly aware of the power of school culture to block change. Nevertheless, to change a culture in school requires an understanding of how it is formed, and how it influences thinking and behaviour. It requires a climate of open discussion about the underlying assumptions of the purposes of being in school, of rules and regulation, and of education as a whole. In the same way, understanding energy efficient buildings and environmental technologies also requires some comprehension of why a reduction in energy demand is needed. The theory underpinning such an approach is the idea that schools are communities rather than institutions, where people construct their own social lives and behaviours rather than have lives created by others.

The Sustainable School? The Educational and Behavioural Context

This is not, however, how the sustainable school is being conceptualized by school leadership teams and the move to more sustainable lifestyles is a priority in few schools. The aim presented by the previous UK government, that every school would be a 'sustainable school' by 2020, has been criticized as overly ambitious (Scott 2009). The educational initiative promoted by the government is the National Sustainable Schools Framework which introduces eight "doorways", activities designed to help schools to operate in a more sustainable way in areas including food and drink; energy and water; travel and traffic; purchasing and waste; building and grounds; inclusion and participation; local wellbeing and global citizenship. It is anticipated, within the framework, that integrated plans across each of these doorways will be developed, but, as Huckle argues (2009), this separation into distinct themes acts to prevent a more holistic educational initiative for sustainable lifestyles. For example, with respect to inclusion and participation, the framework states that by 2020:

> We would like all schools to be models of social inclusion, enabling all pupils to participate fully in school life while instilling a long-lasting respect for human rights, freedoms, cultures and creative expression. (DCSF 2009: 1)

Schools, according to the framework, can promote cohesion within the community and provide an inclusive, welcoming atmosphere that values everyone's participation and contribution, and challenges prejudice and injustice. Secondly, fostering local wellbeing is presented in terms of school pupils exemplifying good citizenship and being empowered by making a difference in their own lives and within their communities:

> We would like all schools to be models of corporate citizenship within their local areas, enriching their educational mission with activities that improve the environment and quality of life of local people. (DCSF 2009: 1)

Third, in terms of global citizenship, the framework puts forward the aim that children should be able to consider the global implications of actions and understand that individuals or countries cannot act in isolation when it comes to reducing carbon emissions:

> The government would like all schools to be models of good global citizenship, enriching their educational mission with active support for the wellbeing of the global environment and community. (DCSF 2009: 1)

But all of these ambitions constitute profoundly difficult educational ideas and demand an examination of complex social and cultural issues. And despite these challenges, a report prepared for the DCSF to examine the impact of sustainable schools activities has presented significant research evidence on how activities have improved the wellbeing of children. The report argues that sustainable schools adopting this framework engage young people, promote healthy lifestyles and make connections with the wider community, thereby enhancing social cohesion (DCSF 2010b). Nevertheless for Scott (2009) as equally for Huckle (2009) the key to achieving a sustainable school relates to developing social capital as much as reducing the use of natural resources. More connected approaches and more critical engagement would allow children to understand the interrelatedness of the "doorways". He posits his own more integrated definition of a sustainable school as one that: manages its use of natural capital to minimize its depletion; has building and equipment which are fit for purpose and as efficient as possible; maximizes human capital by educating people, developing capacity for social action and further learning; and maximizes social capital by adding to social cohesion, wellbeing and mutual understanding, both locally and globally; and teaches about the inter-relationships (Scott 2009).

Thus, different aspects of education policy can act against many of the aims of sustainable development. Whilst the framework for sustainable schools extends the schools' commitment to include care for people across time and space in a global perspective, the current drive for individualism through testing and competition erodes this very principle of care (Huckle 2010). Huckle argues that in order to achieve an integrated approach to education and sustainability, attention needs to be given to what and how students are taught, how the school campus is managed and led (through exemplary buildings and grounds) and how the school can act as a catalyst for change in the wider community. But he also, importantly, argues that these educational goals are undermined by new buildings that are often far from exemplary in terms of their environmental performance and, furthermore, by parents travelling long distances by car and by the schools themselves eroding their integrated approach by the privatization of school catering and the prevention of the use of locally sourced food (Huckle 2010). These flagrant contradictions within different goals of educational policy are noted not only by academic educationalists but also by children themselves (Wheeler 2009). These contradictions between existing flawed educational policy for sustainable development and the culture and behaviour of schools demonstrate the necessity of more holistic approaches to the human, lifestyle and behavioural factors necessary to achieving sustainable schools.

Nevertheless, leadership to change a school culture, one of the recommendations of the Zero Carbon Taskforce, in itself is not easy to put into action. One of the problems of the sustainable schools initiative and the zero-carbon ambition is an educational context poor on definition, and a lack of educational purpose in government policy to create new schools (Biesta 2009). The educational policy context has been characterized by both a rise in the use of spatial language and a shift in emphasis in educational thinking, as evidenced in the emergence of the language of 'environments for learning' or 'learning spaces' (Biesta 2009). Biesta views this development as the result of a shift in educational thinking from the activities of the teacher to the activities of the student, and a change in the role of the teacher to that of a facilitator of the learning process. But whilst Biesta suggests that there is much to be said in support of this shift, there are also consequences with respect to the purpose of education. As he stated in a recorded lecture at the University of Nottingham: '...it is, after all, one thing to create environments that support learning, but it is another thing to create environments that support a particular kind of learning' (2009).

A similar shift from teacher and teaching to pupil and learning is evident in the language of the sustainable school. Participation is a key element in the approach but, as Scott (2010) argues, there is a limitation on what can be asked of children and on the effectiveness of their own agency especially where there are conflicts with adult agendas. In the sustainable schools approach we need to go further and ask even more fundamental questions about the nature of learning.

In the following half of this chapter, I present empirical work with children in schools exploring the notions of sustainable architecture, sustainable lifestyles and education. I argue that there are three aspects to encouraging sustainable living in the school environment: critical engagement; innovative pedagogies; and academic or "green" leadership. Each of these has to be developed before beginning to build sustainable schools. My own work relies heavily on practices of 'co-research' with children and a broadly participative action research methodology and I argue it is these sorts of approaches that can push forward research in school design.

Participation is a legal obligation in the new school context, but significant efforts are needed by both teachers and architects to include children without being merely tokenistic. There are also significant challenges in establishing evidence on which to claim successful participation or the impacts of participation. Participation practices require researchers to rethink their more traditional relationship to the subjects of their research. Cahill (2007) suggests that central to the ethics of participation is a 'retreat from the stance of dispassion' and an 'ethics of care. Including children explicitly in initiatives for change allows researchers to examine how children both experience and constitute school communities. In the setting of the school, participatory research starts with an understanding that children hold in-depth knowledge about their own lives and experiences which should be drawn on to frame the issues and interpretations of research. As Hart and others have proposed, childhood comprises our greatest period of geographical exploration and as such children can make a unique contribution to community development because of their innate knowledge of and feeling for places (Hart 1997, Chalwa 2007).

Children, Consumerism and Consumption

Children negotiate complex social tensions proffered by the commercial world in the context of schools, which can accentuate inequalities and place pressure on those who are already disadvantaged (DCSF 2010b). Children are increasingly targeted by commercial companies in their advertising but evidence of the risks, harm and benefits caused by the commercial world, is rarely conclusive. A report on the impact of the commercial world on children's wellbeing suggests that children are neither the helpless victims imagined by some campaigners nor autonomous and 'savvy' consumers (DCSF 2010b). In the drive towards more sustainable schools, Huckle (2010) proposes the need for change that will impact on schools. Tim Jackson's work on consumerist society mirrors Huckle's argument, where Jackson writes: 'Nature and structure combine here to lock us firmly into the iron cage of consumerism ... we need to identify opportunities for change within society – changes in values, changes in lifestyles, changes in social structure that will free us from the damaging social logic of consumerism' (Jackson 2005: 102). Huckle argues that this economic system would involve the 'co-ordinated participatory planning of production to meet human needs, using appropriate technology, within ecological limits (economic democracy); public/community rather than private provision of many goods and services (eco-localism); useful work for all and guaranteed social wage; reduced working hours allowing time for self-development and community participation; and new forms of governance and democracy – environmental, ecological, global citizenship' (Huckle 2010: 10).

Whilst Huckle argues for a different social, political and economic and structure and Kindon et al. (2007) promote the value of alternative ontologies, others offer a more critical perspective upon commercialism influenced by feminist theory and the inclusion of the hidden voices and calls for new forms of desire (Soper and Thomas 2009). The epistemological position that has influenced my own research is one that attempts to recognize emerging understandings of our relationship to the world and others (Wheeler 2008a, 2008b, 2008c). Pedagogies of connection refer to a critical re-assessment of co-research ethics. Workshops demonstrate children's need to develop a relationship to the world around them and often they express difficult feelings about conflicting adult values. Allowing them to reflect on the notion of place and relationships and encouraging the use of narrative is a means both of sharing these questions with adults and peers and an opportunity to critically reformulate relationships.

What is evident from the following extract of dialogue from children, all of whom were 12 years old at the time, is the ease with which they pick up the concerns of adults but apply the fears, imagination and moral codes of childhood:

Pupil 1: Has anyone seen that movie? *The Day after Tomorrow*?

Pupil 2: Yes

Pupil 1: Some people think that that is going to happen, *The Day after Tomorrow*.

Pupil 3: Oh is that the one where the earth gets flooded? Yes, the world all gets flooded and stuff like that.

Pupil 4: I gave all my clothes to the tsunami when that happened.

Pupil 3: What do you wear then?

Pupil 2: I don't know what's going to happen to the world, who knows what's going to really happen. Whether we're going to get finished off by flooding, whether it's going to fly into the Sun, whether we're all going to die due to global warming.

Pupil 3: We've got a few years left.

Pupil 2: Whether the magma's going to come out and flood the world with magma. Who knows whether someone will create a zombie virus and bring zombies, dead people back to life? Who knows if aliens don't exist and they might destroy the earth. I'm just coming up with theories about what might happen to the earth. I'm thinking we might implode.

These sorts of narratives about climate change in the media were a common cause of questioning, and expressed a certain degree of anxiety. However, stories also emerged reinforcing more immediate concerns and led to wider discussion. A different group of students from the same school were concerned about littering and the unnecessary resources used in packaging products:

Pupil 1: There's a lot of rubbish on the field, more bins around the back for the school.

Pupil 2: Supermarkets are saying to people [to recycle], but they put drinks in packets and wrappers.

Pupil 3: On some packing it says you can recycle it, but some people just chuck it on the floor.

Pupil 2: Because some games, computer games, there's like plastic and you've got to separate it ... they should make an easier way to recycle.

Pupil 3: It's not just like the public getting it wrong because the Government aren't really doing much about it.

Pupil 2: Everyone is just worrying about the credit crunch, at the moment.

Pupil 3: It might be about the public, but it is the Government as well.

In other words, more immediate complaints developed into discussions on ethical manufacture, consumption and the global economic crisis. The need to contextualize responses to questions even where reflecting on the social and economic dimensions of sustainable development was evident. In response to this discussion, and to urge a deeper investigation, the researcher asked whether pupils felt the economic crisis was related to global warming. The response was a narrative which compared personal irresponsibility in money management with the irresponsibility of the banking sector and the social norms of consumerist society:

Pupils: Yeah [all boys responding to the question].

Pupil 1: Because the banks are lending money, but people aren't paying it back...

Pupil 2: Because it's like [a man] maxed out like six credit cards and killed himself, and then his wife had to pay it off.

Pupil 1: Because like if money's gone out of your bank account you won't have enough money to buy light bulbs.

Pupil 2: People want, want, want, they want to go on holidays, they want big cars, they want their children to have the latest video games.

To suggest a question about the firm hold of these social norms, the researcher asked the same group of pupils whether people could stop behaving like this:

Pupil 1: Some kids get spoilt a bit sometimes ... because kids get spoilt my Dad started saying things I don't need and what I want I have to buy it myself. It teaches me how it's going to be like when I grow up. You're limited in what you can buy. And ones that get spoilt should do it as well ... because when they're older it's not going to happen and you need to work for it.

In other words, children understood how behaviour is reinforced by parents and by education. Children's understanding of sustainability merges the personal with the bigger picture in a mix of places and relationships, but in essence the dialogue illustrates just how well children understand and can negotiate personal desire with the pressure of social norms. To demonstrate further, one female pupil in the same group dialogue described how the real root of the problem was 'bad' habit:

> Pupil 3: Is it about habits? It takes a lot to break habits … you know with the green umm… thing it's the way you've been brought up, I think, and the way you act. If you act like you share all the time, you won't be greedy, but if you don't share and you say "no I want that now" not later, that's just greed.
>
> Pupil 1: And if you want it, it's better for like the credit crunch and everything, and it's cheaper, a week later.

In making this comment she suggests that 'sharing' alleviates consumerist desire. To push the conversation further, the researcher asked whether parents' behaviours prevented pupils from adopting more concern for their environments:

> Pupil 1: Yeah, they might.
>
> Pupil 2: Depends on their attitude.
>
> Pupil 1: I want to say that it doesn't depend much on the adults, it's like you act, you don't have to copy them. You can just say "no", "not doing that".

Despite many complaints about the inconsistency of adult behaviours, especially concerning energy efficiency, in this instance, the pupils recognized their own agency. Working within, engaging with, and challenging the school culture with children, is central to effecting change (Sanoff 2003). Children themselves understand and can articulate the potential significance of values as barriers to change, but they also understand that there is a bigger problem to address and, in some instances, that they have personal agency. For example, in a dialogue with another group of students, in this case, the pupils were girls, it was not adults that were to blame, but growth in consumerist societies and globalization:

> Pupil 1: It's not all about schools polluting. There should be a better way to make people stop polluting because scientists and people like that spread their message but what have they done: nothing! So I think school could be more eco-friendly which B***School [the speculative designs for an eco-school were given a name by the participants] are eco-friendly schools but I think that they should in America and Beijing, I'm not saying it's just their fault, but they should cut down on the amount of cars they have because of the Olympics they cut down all the cars they had but now they are reusing all the cars!

In making this comment, the girls perceived those promoting personal action (which was an unwarranted criticism of the researchers but an accurate description for how sustainable schools are being taught in schools) and as neglecting the global dimension to behavioural change. Why Biesta's philosophy of education and his discussion of a "worldly" school are significant in this context, is due to the connection that he makes between the practice of inclusion, education for sustainability and the question of the individual. According to Biesta, a school should be a "worldly" place and education should not be '... just about the transmission of knowledge, skills and values, but ... concerned with the individuality, subjectivity, or personhood of the student, with their 'coming into the world' as unique, singular beings' (Biesta 2006: 27). This humanistic and democratic vision puts emphasis on educational relationships, "on trust, and on responsibility, while acknowledging the inherently difficult character of education" (Biesta 2006: 15). He suggests that the learning environment of the school should motivate children to ask what it means to be human and how we live with others; and not give answers, but incite questions. Teachers have a responsibility for the coming into being of the unique subjectivity of each student: the teacher is responsible for what and who is to come, but without knowledge of what and who is to come.

Participation in the design of their environment and the maintenance and the improvement of the design of existing schools provides an opportunity for young people to explore all these questions and to examine why our relationship with the natural environment has to change. Theorists and researchers in architecture and building engineering tend to separate issues relating to behaviour and technical innovation, but being able to approach complex social and political issues is crucial to effectively achieving low carbon and sustainable design. What presents the challenge is the changing role of architects and other building professionals, the need for architects to be able to act not only as experts in low carbon and sustainable design, but as agents for change; as intermediaries in fostering different sort of relationship to the natural and social environment.

The extracts of dialogue with children presented here illustrate only a sample of the questions about sustainable development that children want to raise. Participatory research conducted with children in post occupancy work demonstrates the need for real dialogue about design with children, about their own experience of school and its places, about their identity and about sustainable development. In recent research we have developed a method for engaging children in post occupancy assessment with the same attention to children's engagement and the spatial experience of the new architecture is often shocking and revealing experiences that demonstrate the real anxiety some children feel. For example, a question about a common circulation and traffic management problem for the spatial design of school and whether it was difficult or frightening when the classes change over using the stairs and it gets crowded resulted in a narrative about the danger posed by the grand stairs rising three floors within the new school building:

Pupil 1: It's hard especially at home time when you are on the top floor and you have to get down all the stairs, especially the main stairs.

Pupil 1: Everyone pushes you out of the way ... and it takes you about 10 minutes to get out and you have to try to hold onto the handrails to pull yourself forward ... I go down with my brother and he makes a little circle and I walk.

Pupil 1: Older people think they are cocky and they can do everything and so they go down the wrong side of the stairs.

Even the need for a social space and time to get away from the 'madness' of others, was discussed. Amongst another group of girls this was illustrated by a space they had found:

Pupil 2: We like to sit under the stairs where there is carpet and a radiator, but we're not allowed. We just like to sit there because it is inside. We just like having a quieter area you can sit and just be with your friends.

Wellbeing and Sustainable Schools: Developing Pedagogies of Connection

The importance of environmental experience on children's physical and mental health and wellbeing has been established (Hart and Chalwa 1982, Chalwa and Cushing 2007). If children are to be encouraged to make lifestyle changes they will have to enter and be included in a discussion of our relationship to the environment, to community and social cohesion, and all of the political and philosophical complexities this entails. Young people will have to confront and reconcile conflicting social pressures. Place, identity and wellbeing are key to newly forming identities. Reducing carbon emissions requires a human response, not only a scientific response, and changing human behaviours will be an ongoing learning process. Children need ways to discover and continuously reformulate their relationship to the world and others. This will require more than just 'spaces for learning' but schools whose culture and ethos reflect the need for change and architectural environments that are designed to reflect a requirement for technological innovation and behavioural change.

Involving children in design provides opportunities for children to examine the social and cultural factors impeding the adoption of more sustainable lifestyles. The research demonstrated the ease with which children can be inspired to explore the problem of sustainable development and energy behaviours. Narrative, performance and speculative design were illustrative of this engagement. This contradicts common understanding about a lack of both knowledge and engagement amongst young people. Research processes have therefore allowed children to explore more authentic relationships to the environment, to engage with some of the wider social and political problems of sustainability and to discover their own relationship to the world and to others. Pedagogies of connection describe the processes involved, but if architects or other construction professionals take on a much-needed role of agents of change, the demand placed on them, and the skills required, cannot be underestimated.

References

Biesta, G. 2009. *Creating Spaces for Learning or Making Room for Education? Transforming Our Schools* Lecture series, The University of Nottingham.

Biesta, G. 2006. *Beyond learning: Democratic Education for a Human Future.* Boulder, CO: Paradigm Publishers.

Blair, T. 2004. *Speech by Prime Minister Tony Blair on the 10th Anniversary of the Prince of Wales' Business and Environment Programme* 14th September 2004 [online]. Available at: http://webarchive.nationalarchives.gov. uk/20061023193551/number10.gov.uk/page5087 [accessed: 14 November 2010].

Bordass, W. 2009. *Passivhaus Schools: The Route to Low-Energy Schools in the UK?* [Online]. Available at: http://www.usablebuildings.co.uk/Pages/ UBEvents.html [accessed: 14 November 2010].

Bunn, R. 2009. *Sustainable Schools: Defining the Issues* [Online]. Available at: http://www.usablebuildings.co.uk/Pages/UBEvents.html [accessed: 14 November 2010].

Cahill, C. 2007. The personal is political: Developing new subjectivities through participatory action research. *Gender, Place & Culture*, 14(3), 267-92.

Chawla, L. and Cushing, D.F. 2007. Education for strategic environmental behaviour, *Environmental Education Research*, 13(4), 437-52.

Chawla, L. 2007. Childhood Experiences Associated with Care for the Natural World: A Theoretical Framework for Empirical Results. *Children, Youth and Environments*, 17(4), 144-70.

Collier, A., Mackle, R. and Pike, T. 2010. *Applying 'Nudge'*. Paper to Green Nudges. Individual behaviours, social influence and environmental sustainability. Manchester Museum, UK, June 2010.

Department for Children, Schools and Families. 2010a. *The Road to Zero Carbon: The Final Report of the Zero Carbon Schools Taskforce.*

Department for Children, Schools and Families. 2010b. *Evidence for Impact of Sustainable Schools.* Nottingham: DCSF.

Department for Children, Schools and Families. 2009. *National Framework for Sustainable Schools Poster* [Online] at: www.esd.escalate.ac.uk/ downloads/1438.pdf [accessed: 14 September 2011].

Department for Education and Schools. 2004. *Every Child Matters: Change for Children.* Nottingham: DFES.

Gill, Z.M., Tierney, M.J., Pegg, I.M. and Allan, N. 2010. Low-energy dwellings: The contribution of behaviours to actual performance. *Building Research and Information*, 38(5), 491-508.

Gram-Hansen, K. 2010. Residential heat comfort practices: Understanding users. *Building Research and Information*, 38(20), 175-86.

Hart, R.A. 1997. *Children's Participation: The Theory and Practice of Involving Young Citizens in Community Development and Environmental Care*. New York: UNICEF and London: Earthscan.

Huckle, J. 2009. Sustainable Schools, responding to new challenges and opportunities *Geography* [Online]. http://john.huckle.org.uk/publications_downloads.jsp [accessed: 14 November 2010].

Huckle, J. 2010. Consuming Less and Aspiring Differently. *School Design Futures*, [Online] Available at: www.ukerc.ac.uk/support/tiki-index.php?page=1004_MP_SchoolDesignFutures [accessed: 14 November 2010].

Jackson, T. 2005. *Prosperity without Growth*. London: Earthscan.

Kindon, S., Pain, R. and Kesby, M. 2007. *Participatory Action Research Approaches and Methods: Connecting People, Participation and Place*. London: Routledge.

Newman, M., Woodcock, A. and Dunham, P. 2009. Results from a Post-Occupancy Evaluation of Five Primary Schools, in *Contemporary Ergonomics*, edited by P.D. Bust. London: Taylor & Francis, 462-70.

Sanoff, H. 1992. *Integrating Programming, Evaluation and Participation in Design*. Brookfield, VT: Avebury.

Sanoff, H. 2003. *School Building Assessment Methods*. National Clearinghouse for Educational Facilities [Online]. Available at: http://www.edfacilities.org/pubs/sanoffassess.pdf [accessed: 14 November 2010].

Scott, W. 2009. Critiquing the Idea of a Sustainable School as a Model and Catalyst for Change. *Transforming Our Schools* [Lecture series]. The University of Nottingham.

Scott, W. 2010. The Sustainable School: Examining assumptions and young people's motivations, interests and knowledge *School Design Futures* [Online]. Available at: www.ukerc.ac.uk/support/tiki-index.php?page=1004_MP_SchoolDesignFutures [accessed: 14 November 2010].

Soper, K. and Thomas L. 2009. *The Politics and Pleasures of Consuming Differently*. London: Palgrave.

Stevenson, F. 2008. *Post-occupancy Evaluation of Housing* [Online]. Available at: www.usablebuildings.co.uk/Pages/UBEvents.html [accessed: 14 November 2010].

Stevenson, F. 2010. Preface: In this special issue. *Building Research & Information* 38(5), 435-6.

Wheeler, A. 2008a. Architectural Issues in Building Community through Luce Irigaray's Perspective on being-two, in *Teaching*, edited by L. Irigaray. London: Continuum, 61-8.

Wheeler, A. 2008b. About being-two in an architectural perspective, in *Conversations*, edited by L. Irigaray. London: Continuum, 53-72.

Wheeler, A. 2008c. Building Sustainable Schools: Are places of social interaction more important than classrooms? in *Exploring Avenues to Interdisciplinary Research: From Cross- to Multi- to Inter-disciplinarity*, edited by M. Karanika-Murray and R. Wiesemes. Nottingham: Nottingham University Press.

Wheeler, A. 2009. The Ethical Dilemma of Lifestyle Change: Designing for sustainable schools and sustainable citizenship. *Les ateliers de l'éthique*, 4(1), 140-55.

Chapter 8

Is 'Modern Culture' Bad for Our Wellbeing? Views from 'Elite' and 'Excluded' Scotland

Sandra Carlisle, Phil Hanlon, David Reilly, Andrew Lyon and
Gregor Henderson

The orientation of this chapter comes from a broadly public health perspective, but one with a deep interest in how 'modern culture' influences wellbeing, for good or ill. Funded by the Scottish Government's National Programme for Improving Positive Mental Health and Wellbeing, we investigated this relationship through an exploration of just some of many relevant multiple literatures and intellectual disciplines (Carlisle and Hanlon 2007a). We then augmented this (inevitably partial and limited) attempt at theoretical synthesis with qualitative fieldwork in Scotland. This chapter presents findings from both strands of our enquiry. In the first section below we provide a brief outline of the ways in which the rapidly growing field of wellbeing research, which is largely based within the disciplines of psychology and economics, focuses mainly on the psychological (cognitive and emotional) dimensions of wellbeing and, to a lesser extent, its social-structural dimensions. We then sketch some key thinking from other disciplines which suggest that certain contemporary cultural traits associated with modernity, such as materialism, consumerism and individualism, may be profoundly damaging to wellbeing at individual and social levels.

The issues briefly outlined above may well be of considerable interest to the academic community, but they also prompt the question of whether a culturally-informed critique of late modernity has any resonance beyond academia, and with lived experience. In the second part of the chapter, therefore, we describe how we investigated these issues with people living and working in Scotland. We initiated conversations with groups and individuals from 'Elite' and 'Excluded' Scotland: these terms are not intended as descriptions of people or social groups – and they are clearly not descriptions of any geo-physical location. Rather, these are concepts that help capture the sense that Scotland may be a single country, but it is also a society which occupies a range of social, economic and cultural spaces – observably different 'nations', in the terminology of the Scottish Council Foundation (McCormick and Leicester 1988).

We suggest that our empirical work puts some flesh on the bones of the cultural critique surrounding wellbeing in modern society. However, thinking from a third research strand is necessary to deepen our understanding of the globally interconnected nature of emerging problems of wellbeing and place. Evidence from the environmental sciences and thinking from philosophy have a vital place here. We therefore conclude our chapter by considering the implications of Western-type consumer culture, and its apparently endless fixation with continued economic growth, for human wellbeing in the context of emerging global challenges such as climate change, peak oil and economic crisis.

Wellbeing Research: Complex and Contested Territory

Research from economics and psychology tells us that Western societies have, until very recently, experienced decades of sustained economic growth which has resulted in great improvements to health and social conditions. The apparent paradox is that such affluent societies have also seen static or even declining levels of mental health and emotional wellbeing (Easterbrook 2004, Easterlin 1974, Lane 2000, Layard 2006). Rising forms of mental disorder have emerged over roughly the same time frame (James 2007, Schwartz 2000, World Health Organization 2001). On the one hand, the psychology of wellbeing – and the new discipline of Positive Psychology in particular – provides much evidence about a variety of individualized techniques that can be used to improve our mental health and wellbeing, at least in the short term (Seligman and Csikszentmihalyi 2000). Knowledge about how positive feelings and positive mental functioning can be enhanced comprises an important component of such research. Concepts such as 'the pleasant life', 'the good life' and 'the meaningful life' have been used to structure different aspects or levels of wellbeing (Nettle 2005). At the same time, a number of commentators have expressed their critical wariness of creeping tendencies towards ever-more intrusive forms and modes of therapeutic surveillance and self-surveillance in contemporary society (Furedi 2004, Horwitz and Wakefield 2007, Sointu 2005, Williams 2000). These are just a few examples drawn from the large, complex and rapidly increasing bodies of literature around wellbeing, which begs the question, what can be added by focusing on 'culture'?

Though firm definitions remain elusive,[1] 'culture' can be understood as the learned system of negotiated meanings and symbols which frame the way people perceive the world, shape the identities they are obliged to forge for themselves within that world, and act in it. To speak of the importance of culture is to insist on the need for understanding the place of meaning and significance in all human life (Geertz 1973). From this perspective, culture is not simply an external factor or an internally-experienced phenomenon: it is the means by which, and the space in which, we routinely make sense of life, and act on the basis of that sense-making process. The cultural spaces of life are often taken for granted and misrecognized as part of the natural order of things, rather than a human construct. Yet culture clearly matters in relation to wellbeing because it influences the ways in which we

---

1　Geertz (1973) discerns about 200 different meanings attached to the word culture listed in a seminal anthropological work, Clyde Kluckhohn's (1949) *Mirror for Man.*

understand the world and the goals in life which we value and pursue. Moreover, as culture can also influence the distribution and availability of social, economic and cultural resources (Bourdieu 1984, Offer 2006, Wilkinson 1996) necessary to attain such goals, it has considerable relevance for our understanding of how social inequalities are created and perpetuated (Carlisle et al. 2008).

Eckersley's (2005) synthesis of much research on the stagnation or decline of wellbeing in contemporary society suggests that 'modern' cultural values such as economism, materialism, consumerism, and individualism may be responsible. Economism is the tendency to view the world through the lens of economics, to view a society principally as an economy, and to believe that economic values are the most important for our individual and social wellbeing. Materialism and consumerism involve the attempt to acquire meaning, happiness and fulfilment through the acquisition and possession of material things: non-material aspects of life may be squeezed out. And where individualism has high cultural value, the onus of success in life rests with the individual, who is subject to the tyranny of excessive choice and higher expectations, whilst experiencing less social support. Eckersley's research also suggests that many people living in 'modern' society suffer from a lack of purpose and meaning.

Similar conclusions are also found in the psychological, philosophical, sociological and political science literatures (e.g. Czikzentmihalyi 2004, Hartmut 1998, Lury 2003, Schwartz 2000, Sennett, 2006, Slater 1997). Bauman, one of the foremost theorists of modernity, has noted that the dominance of an individualized consumer culture, in the context of a neo-liberal and market-driven society, results in a heightened sense of risk, anxiety, uncertainty and insecurity for *everyone* but most particularly for the 'new poor', a category of people he tellingly describes as 'flawed consumers' (Bauman 1998, 2001). A key function of this disadvantaged group is to serve as a warning to the rest of us! He is echoed in this by political scientist Rosa Hartmut (1998), who has suggested that our life choices are far more restricted than we realize, and are largely structured to serve the workings of the modern capitalist economy.

In a chapter, which briefly reviewed some of the literatures referred to above, we suggested that many people are perhaps too comfortable to want to change, even if consumer culture is depriving their lives of real purpose and meaning, whilst more disadvantaged groups and individuals may lack the resources for change (Carlisle and Hanlon 2007b). But we also wanted to know whether this diagnosis of cultural 'dis-ease' resonated more widely, rather than simply being a concern of academia. So we turn at this point to the ways in which our empirical, place-based findings bear on such research.

'Soundings' From Scotland

Through fieldwork conducted in Scotland we sought to explore whether people perceived 'modern' culture as a significant factor in wellbeing, and if so, how they articulated and responded to such trends. But first we might ask 'why Scotland?' In other words, what is the relevance of place to experience in this context? One of the cradles of modernity, Scotland has undergone a series of rapid socio-economic transitions in the comparatively recent past, including that from industrialization to post-industrialization. This country has seen large scale unemployment, poverty and some of the worst health outcomes in the developed world for some of its population (Leyland et al. 2007), whilst others have flourished. Ascherson (2003) has designated this division as the 'St Andrew's Rift'. The health problems which are now overtaking diseases with marked structural/material correlations (such as Coronary Heart Disease, stroke and cancer) are arguably cultural as much as structural/material in origin and nature, such as obesity, problematic alcohol and drug use, and the physical results of violent assault (Hanlon et al. 2006). Rates in anti-depressant prescribing are rising and suicide rates are higher than in other parts of the UK (Platt et al. 2005). This pattern of emerging problems suggests that individual and social wellbeing may possibly be under greater threat in Scotland than in comparable places elsewhere (Craig 2003, Walsh et al. 2010). Work by the Scottish Council Foundation (McCormick and Leicester 1988) has also drawn attention to the divided nature of Scottish society and the considerable socio-economic differences that exist between what they designate as the 'Three Scotlands'. In sum, Scotland might well be a suitable place in which to explore questions of cultural space.

Methodology

For the purposes of our research we wished to explore questions of if – and how – the cultural spaces of modernity also shape people's experiences across two[2] of the otherwise very different social spaces postulated by McCormick and Leicester (1988). We interviewed twenty-eight people across eight different settings, in two major Scottish cities and their environs, over three days in late 2006/early 2007. Our sample was purposefully selected as illustrative of the extreme diversity of Scottish life and contained people able to speak from a wide range of perspectives and experiences. At the affluent end of the socio-economic spectrum we interviewed elite and influential people working at senior levels within both public and private sector settings ('places', in the conventional sense). Views from the other end of the Scottish socio-economic spectrum were sought from participants in a community project based in a deprived urban neighbourhood, and a group of prisoners. The latter two groups can reasonably be characterized as disadvantaged and excluded. Box 8.1 is based on McCormick and Leicester's conceptual schema of 'three nations' in Scotland, although only 'elite' and 'excluded' Scotland are represented here.

---

2   Although we focus here on the contrasts between 'elite' and 'excluded' Scotland, we report elsewhere on findings from a (mainly public sector) middle grouping ('secure Scotland') in an otherwise tripartite conceptual schema (Hanlon and Carlisle 2009).

Box 8.1   Elite and Excluded Scotland

Elite Scotland is a reasonably secure place to live – not risk-free, but people have the resources to insure themselves against risk. Household incomes are secure and firmly upwards. Elite Scotland is mostly affluent, well housed, healthy and mobile. While working long hours is a common feature, those living in Elite Scotland also have the money to buy enough time to enjoy leisure activities. The gap between incomes in working age and retirement is the smallest of any group in Scotland. Elite Scotland is made up of households with most choice. They are financially the best equipped to cope with rapid changes of circumstance.

Excluded Scotland is an expensive place to live, where household basics consume most of the family budget. The risk of long-term poverty and social exclusion is highest. Households are more likely to live in poorly-served housing estates as well as poorly maintained private housing. The risk of homelessness is greatest. Households are most likely to have no earners and stay on benefits for longest. Poor households in Excluded Scotland are more likely to live in poor communities: they are poorly networked. The risk of isolation and depression is high, for lone parents, older people and victims of racial harassment and other types of crime. Households are most exposed to market failure, monopolies and lack of choice. Many are cash-poor but time-rich and have other skills to offer to their communities.

One member of the research team (AL) organized and facilitated our visits. Each research team member was present during at least one whole day of interviewing and all took an active part in developing discussions with our participants. Each member was also involved in subsequent processes of discussion, analysis and developing consensus around our findings. In each visit our study was introduced and discussions initiated by one of the researchers (PH), who provided a verbal summary of the cultural critique outlined above as an initial prompt for discussion. Another member of the group (DR) took responsibility for asking questions. Others within the research team contributed freely to the discussion as appropriate.

Our interviews were not tape recorded as some of our interviews were conducted in places where this was not possible, either because of the sheer number of participants or because individuals and groups were physically moving around. All members of the research group made notes for later comparison and one of the researchers (SC) used shorthand to take more detailed notes and capture much verbatim comment. These shorthand notes were then transcribed as the main basis for analysis, in conjunction with notes made by other team members, as the writing of field notes is not simply a craft skill but also a key part of the analytical process (Emmerson et al. 1995). This collaborative process enabled us to have confidence that most, if not all, elements of our discussions were recorded with reasonable completeness. Analysis took the form of several iterations and refinements of the data provided by our notes. Themes which emerged in the particular context of each visit were highlighted in subsequent iterations of the fieldwork account. A final iteration took the form of an anonymized report on each visit, which was then sent to our host groups, to ascertain whether our account matched their interpretations. Participants were then invited to confirm this or suggest amendments: any amendments made were minor in nature.

The following two sections present our findings in terms of 'soundings' taken from 'Elite' *versus* 'Excluded' Scotland. We use the term 'soundings' to indicate the tentative nature of our conclusions. Inevitably, lengthy conversations have had to be compressed into a small number of pertinent verbatim quotations which illustrate the main themes.

Soundings from 'Elite' Scotland

Wellbeing in Public Life: Times are Hard for Dreamers[3]

From senior academic staff at a renowned art school we expected to hear about the capacity for creativity and autonomy to be found in their world, which might help create and sustain a sense of wellbeing. And indeed our respondents spoke about a professional world that seemed, on the positive side, to be characterized by flexibility, responsiveness, and integration of diverse creative and artistic processes towards shared goals. Yet they also told us that creativity is now struggling with conditions of unprecedented technological change and complexity: their students also seemed driven by primarily instrumental aims. The uncertainty of contemporary professional life, within the accountability-driven domain of the public sector, emerged during our conversations. We heard that:

> any sense of the spiritual is gradually being squeezed out

and that:

> human qualities are being subverted by the instrumental agenda currently dominating public life.

This theme of instrumentalism, of times being hard for visionaries, was strongly echoed in our conversations with the Headteacher of a secondary school who observed that:

> people everywhere live busy, consumer-driven lives. Their energies are devoted to money, mortgages ... there's no time for emotional energy.

---

3   This quotation is from graffiti found on the wall of the women's toilet in the art school.

This is reflected in his local area, he said, where parents are mostly public sector and lower level commercial workers and:

> want to *be* something – they're conscious of material wealth and income and have worked hard to get here.

His school was sited in an aspiring middle-class area, but he believed that its inhabitants still experienced poverty:

> they have more stuff, cars, washing machines – but they're still dreadfully poor, because it's relative.

In such circumstances, education becomes a consumer product. He observed that local parents, perhaps because of their own insecurities, can bring an aggressive approach to their child's education:

> they know that educational qualifications are what allow you to make that step up. So they push for results. They can't afford to pay the fees but they want to treat us as a private school. There's a sense of consumerism there – we have a product to offer and we'd better deliver the goods!

Their children apparently share these views:

> it's very much "I want to be a lawyer, doctor or dentist". They're very materially focused.

Pupils' main interests lie in:

> being able to get stuff from the classroom that will lead to qualifications.

He judged that there are no opportunities in contemporary school life for young people to develop "an inner voice", as there was:

> a poverty of vision holding people back. That and the false beliefs people hold.

Empowering The Workers: Wellbeing As Good Business Sense In Corporate Culture

The fast-paced, results-oriented business environment of a city call centre set up to service a large manufacturing industry provided a very different setting in which to explore our research questions with others. We heard how a Human Resources Director of this private sector company sought to find ways of combating the dehumanizing tendencies of a call centre environment. He told us that

taking calls from people moaning all day and working from a pre-designed
script results in de-motivating work, low staff morale and high staff turnover.

His approach involved giving staff intensive training plus some autonomy in
addressing the problems and complaints of customers:

giving people the tools and freedoms to deliver better business practices ...
training people to think positively and look for solutions.

These techniques enabled the company to achieve low rates of staff loss and high
levels of customer satisfaction. This mattered because, although treating workers
with respect and investing their work with dignity was regarded as important, his
"hippy culture" was:

no use unless I can deliver the business – failure would be dreadful – it would
confirm that the fluffy stuff doesn't work!

Within the obvious limitations of the call centre environment, efforts to imbue
otherwise deadening work with creativity and value, were being made. However,
although the call centre workers worked hard, they played hard too. We heard that
Friday night binge-drinking, in the context of a city with a vibrant club culture, was
an entrenched part of young workers' lives and one on which company policies had
little impact.

In our second visit to a private sector organization, a high-tech communications
business, we heard from the General Manager that his company seeks to give its
employees the tools to manage the work-life balance through flexible working
technology. This, we heard, reflected the culture of an organization which seeks
to support vulnerable employees, such as those with caring responsibilities. He
acknowledged that his was a massively competitive environment in which to work
and subject to constant change and innovation. In this environment, 'how you
treat people, and their health, matters'. Organizational cultural change was partly
driven by cost considerations, we were told, but also by:

the genuine belief that it is in both the individual and corporate interest to behave
in socially sustainable ways.

These investments paid off in that skilled and experienced staff were retained.
However, wellbeing in this organization had to be balanced with corporate
pressures to do better, faster: as employee stress levels appeared to be rising, the
problem had not been solved.

He also reflected on the influence of consumerism in his own personal and
professional life, which he tried to resist (not least because his senior role released
him from the obligation to 'push product'). Yet a profound and disturbing gulf

of understanding between elite and disadvantaged people is suggested in his comment that:

> life is easier for parents who can't afford to give their kids stuff. It's harder for those who have money to say no, when they can pay.

In Defence of the Market Economy

Perhaps not surprisingly, senior representatives of consumer and industry organizations suggested to us that consumerism is merely a superficial manifestation of deeper structural and cultural problems in our society and not, in itself, the main problem. Our respondents acknowledged that the market impacts differently on rich and poor because:

> whilst poorer people aspire to demonstrate what they've got, the wealthy don't need to.

They argued that:

> the market economy *is* the prevailing culture and ideology of the modern world, including Scotland.

As alternative social models tend to be coercive, and corrosive of freedom, there really is:

> no alternative to market capitalism. All we can do is find mechanisms to mitigate the excesses, because people also understand that selfish actions can have social costs.

These participants spoke of their interest in fairness and of their role in protecting disadvantaged consumers. One person acknowledged the need to adjust the 'nasty, brutal, monopolizing force of markets' and spoke of the need for regulatory impact on advertising and the problem of unequal access to information. But they also spoke of the need to balance consumer with producer interests. We heard a defence of the free market system as 'the least worst option'. In response to our questions about the sustainability of current ways of life, one of our hosts remarked that such arguments invariably end up as "choice editing", i.e. denying to people their right to choose. It was, they believed:

> better to make people pay for their choices, rather than restrict them ... a critical stance towards consumerism is an elitist road to go down.

Similarly, it was suggested that those who criticize consumer society risk moving towards a centralizing and undemocratic position, based on the idea of a society where:

> somebody else makes choices for people. That kind of society happens through political choices, and our society is not likely to make them.

One of our hosts remarked that Western ideas of political philosophy are built on certain freedoms which may not sit easily together, for example freedom from want, alongside freedom to make choices or pursue one's own goals. Despite these tensions, we were told that:

> there is no obvious challenge to the underlying consensus in favour of a market-oriented capitalist society.

In this context, they argued, markets remain:

> the most efficient and least unjust way of organizing society, even though this involves great disparities of wealth.

Soundings from 'Excluded' Scotland

Life and Losses in the Consumerist Bubble

The pervasive influence of consumer culture on individual, family, community and social wellbeing was powerfully voiced during our conversation with the group of male prisoners. One of our respondents suggested that contemporary culture influences wellbeing because it is profoundly isolating:

> people live in their own bubble, getting in their own car to drive to work, staying in their own home. Community spirit has gone, and this compounds the issue.

Several members of this group referred to their own parenting experience. One argued that, even in difficult circumstances:

> it's possible to provide a stable background and enjoy a good family life, though this involves some sacrifices.

He concluded that 'parents have let their kids down'. Another man, however, argued that some families need both parents to work in order to survive: he had given his family a good standard of living but had to work away from home in order to do that. One of our hosts suggested that 'happiness is a close family unit' but another responded:

That's gone. We're all in debt. You're stressed, you go to work, you go home. You sit in front of the TV. There's no family dinner, no time to talk problems through, sort things out. You're just working to afford that TV. There's no time for your children when you come home at night. No time to talk.

One of his peers contended that children are also under cultural pressure, not least because of media influence, and that there is now:

a greater divide between the haves and have-nots. And young people also want the big house, flash car and plasma TV.

Another suggested that:

Capitalist society is about the stimulation of insecurity. A community that feels safe, for example, has less need to purchase insurance.

Under such circumstances, people are driven to:

buy products and commodities to solve their perceived problems – such as cosmetic surgery.

Another man spoke of the enormous profits made by the large banking corporations who constantly attempt to get people to borrow money, 'offering credit through junk mail'. He suggested that there is 'real pressure to go into debt in order to have material things', whilst one of his peers countered this with the argument that:

it's wider than just materialism. You want to do well and move on in order to get a better class of partner, a better standing in the community.

One of this group concluded that:

our focus needs to go down to the spiritual – to the value and worth of a human being. Virtually nothing in society promotes that. We are exploitable because we are fearful... if you live in a society that's been founded on exploitation of the masses, how are you going to de-condition them? How do you make people feel more confident in themselves?

When we asked him what would help with this, he responded that we were all:

trapped in a cycle of consumerism. Powerful groups can't be expected to support anything that will counter techniques for maintaining social dominance.

We were told that people within the prison population and those in the housing estates outside tend to cope with the pressures of modern life through drug use.

One man remarked on the presence of formal rites of passage in other cultures, which provide a sense of progression and development, and their absence from our own culture – as in the loss of the apprenticeship model for the non-academically inclined. All our society can provide, he suggested, are:

> a series of transitions – from primary school to secondary school, and from there
> to further education, work, or the dole queue.

Salvage and Sanctuary: Myths, Metaphors and Meaning on Clydeside

The most hopeful account of how alienating social and cultural trends could be resisted emerged during our conversations with workers and participants in a boat building project located in a particularly impoverished neighbourhood on the river Clyde. Here, metaphors about the re-creation of meaning and purpose in damaged lives were given physical form in the practical spaces of a wood yard and workshop. We were received with courtesy by men bearing knife scars as witness to harsh lives, and encountered striking visual and verbal metaphors of storm-felled lives amongst the storm-felled trees from which project members built their boats. The project aimed to connect inner city Scots with their more rural past and also with the strong cultural heritage of boat building on the Clyde. We were told that unemployed people themselves had started the project – building a boat as their way of reconnecting with history and heritage. The project worked with women as well as men, teaching crafts such as weaving, metalwork and other craft skills.

It also looked further back in time, to the ancient past of 9th Century Scotland and the inclusive system of clan relationships. There were marked parallels here with the project's present day aims, one of which was to recreate a sense of community and inclusion. Training and practice of a skilled craft created dignified ways of building a sustainable livelihood in otherwise impoverished circumstances. Aimed mainly at homeless people with various addiction problems, participants were only turned away if uncontrolled drug addicts, even so:

> the door is never permanently closed.

Nor were participants ever required to leave as, once training is complete and membership enduring: 'once you're in the Clan, you're in'. Every participant was believed to possess an intrinsic capacity to contribute to the work. One project participant told us that, in working on the wood to create a boat, each man was also re-shaping his own life. More prosaically, we were told that:

> it's about getting people away from watching rubbish on daytime TV.

One older man who had stood silently listening whilst others talked remarked that:

people are more relaxed here, get treated with respect and give respect back. There are no strict rules.

He said, of people like himself:

you get that look from people, out there on the street. In here you're treated as a human being.

Here, as in the prisoner group, we found arguments that many of us now live in ways that damage not just ourselves but the environment on which we ultimately depend. One of our hosts claimed that 'modern culture destroys people's hearts and spirits', whereas the boat building project provided:

a place where miraculous things can happen. Outside they're trodden down. In here they've always got something to give.

Summary

We found recognition of some malign cultural influences, together with differences in the ways in which participants conceived and responded to these. For example, we discovered a shared acknowledgement that issues of individualism, materialism and the driving force of the market economy are pertinent and forceful in contemporary lives. Nevertheless, we also found that the *meanings* of those issues differed, depending on respondents' social, economic and cultural locations. Our elite research participants tended to engage with the issues we sought to discuss in an abstract, impersonal and professionally contextualized way. Disadvantaged or excluded participants expressed far more personalized and critical views on the nature of modern life and its effects on vulnerable individuals, the family, community and society in general. On the one hand, this may reflect the discursive styles of the different groupings. However, it may also indicate that people experience the influences of materialism and consumerism in different ways, depending on their socio-economic position.

Our sense is that the lives of more affluent participants were almost certainly buffered by virtue of their more advantaged socio-economic position in 'Elite Scotland', together with their possession of substantial forms of economic, educational, social and cultural capital. Conversely, the excluded (and stigmatized) groups we spoke to seemed to have experienced the full force of the more damaging traits of modernity, though the boat building project also exemplified resistance and provided alternative meaning and purpose.

At this point we return to the third research strand referred to in the introduction to this chapter, where we suggested that much research on wellbeing (including qualitative forms of inquiry into lived experience) remains sufficient to help us understand the globally interconnected nature of emerging problems of wellbeing and place.

The Environmental Critique

The effects of global economic crisis and planetary climate change now loom large in many contemporary lives. Though less familiar, the phenomenon of 'peak oil' bids fair to change our society in radical and potentially disruptive ways. This suggests that steadily accumulating evidence from the environmental sciences has a vital place in making connections and deepening our understanding of the relationship between wellbeing and place. The emerging environmental critique successfully draws together evidence and thinking from diverse fields in shared concerns about the consequences of the increasingly dominant – and globalized – cultural system associated with Western modernity.

As one philosopher has noted, the increasing obsession in 'modern' society with superficialities such as wealth, fame, physical appearance and material possessions is linked not just to the decline of care and concern for others in the world, but to the neglect, even potential destruction, of humankind's shared environment (Cafaro 2001). From this perspective, continued over-consumption by the few may ultimately render the physical world uninhabitable for all. The dominant cultural norms and values found in many Western-type societies have resulted in a marked imbalance between a specific way of life and the planet's environmental carrying capacity (Harrison 1993). Humanity faces looming global changes as at least a partial result of this imbalance. These include anthropogenic climate change (IPCC 1996, 2007, McMichael et al. 2006), which could lead to multiple socio-economic impacts such as mass migration and many public health challenges. We may now have passed the peak in oil production (Hubbert 1945, Roberts 2005). The loss of this cheap energy resource on which we have long depended will inevitably lead to dramatic social change on a global scale (Hanlon and McCartney 2008). The cumulative consequences of modern consumerism have resulted in continued three, five or even seven-planet living[4] for the wealthiest nations, with one-planet living for the rest (Marks et al. 2006).

The environmental critique is now, of course, virtually ubiquitous. It is therefore not surprising that psychological researchers Ryan and Deci (2001) warn that, as individuals pursue satisfying and pleasurable aims within affluent societies, they may create conditions preventing the attainment of wellbeing by others. They urge attention to the relationship between personal wellbeing and broader issues such as the collective wellness of humanity and the planet.

---

4   The concept of multi-planet living indicates the level of natural resources that would needed if the rest of the world were to consume at the level of those in the wealthiest societies. Estimates of numbers of planets needed vary, depending on the affluent nation being considered.

In similar vein, Csikszentmihalyi (2004), one of the founders of the Positive Psychology movement, argues that humanity needs a new image of what it means to be human: the efforts of those living in modern societies to create comfortable environments in the belief that this will improve life, have actually undermined the essence of what makes life worth living over the long run. Yet despite this dawning awareness across many disciplines of the interconnected nature of issues such as human wellbeing and global sustainability (and the emergence of much political rhetoric on these topics), tokenistic change seems the best the wealthiest nations have as yet been able to manage (Downing and Ballantyne 2007).

Are Meaningful 'Conclusions' Possible?

This chapter highlights the need to integrate cultural perspectives into our analyses of and interventions around wellbeing. Inevitably, much has been over-simplified or omitted in our brief attempt to synthesize a mass of complex evidence and theory from diverse and often disconnected sources. Nevertheless, one conclusion suggested by our readings of the literature(s) and our limited foray to 'test' such findings through fieldwork in Scotland is that, if it is the case that modern consumer culture threatens wellbeing (at the level of the individual and beyond), then many disciplines and professions would benefit from a better understanding of how this happens. Culture, possibly the unique possession of *Homo sapiens*, shapes the spaces and places in which all humans live, but not all cultures are considered equal. Research indicates that the values of 'modern' culture work to influence wellbeing at the level of the individual, society and, ultimately, the planet on which we all depend. This is because 'modern' culture is not just implicated in the daily (re) creation of contemporary structural and material conditions in the affluent world of the global North but also in the visions of 'the good life' (based on apparently limitless economic growth) exported to and adopted by other, poorer societies as a model of how best to live. Not only is this a demonstrably flawed message, it is simply not sustainable on a global scale. An inescapable conclusion is that the relationship between culture, wellbeing and place are important far beyond both the ivory towers of academia and the grass-roots work of environmentalists: this is an issue which will, sooner or later, impact on all humankind.

References

Ascherson, N. 2003. *Stone Voices: The Search for Scotland.* London: Granta Books.

Bauman, Z. 1998. *Work, Consumerism and the New Poor.* Buckingham: Open University Press.

Bauman, Z. 2001. *The Individualized Society.* Cambridge: Polity Press.

Bourdieu, P. 1984. *Distinction: A Social Critique of the Judgement of Taste.* London: Routledge.

Cafaro, P. 2001. Economic Consumption, Pleasure, and the Good Life. *Journal of Social Philosophy*, 32, 471-86.

Carlisle, S. and Hanlon, P. 2007a. The complex territory of wellbeing: Contestable evidence, contentious theories and speculative conclusions. *Journal of Public Mental Health*, 6(2) 8-13.

Carlisle, S. and Hanlon, P. 2007b. Wellbeing and consumer culture: A different kind of public health problem? *Health Promotion International*, 22, 261-68.

Carlisle, S., Hanlon, P. and Hannah, M. 2008. Status, taste and distinction in consumer culture: Acknowledging the symbolic dimensions of inequality. *Public Health*, 122, 631-37.

Craig, C. 2003. *The Scots' Crisis of Confidence.* Edinburgh: Big Thinking.

Csikszentmihalyi, M. 2004. What we must accomplish in the coming decades. *Zygon*, 39, 359-66.

Downing, P. and Ballantyne, J. 2007. *Tipping Point or Turning Point? Social Marketing & Climate Change.* London: Ipsos MORI Social Research Institute.

Easterbrook, G. 2004. *The Progress Paradox: How Life Gets Better While People Feel Worse.* New York: Random House.

Easterlin, R.A. 1974. Does Economic Growth Improve the Human Lot?, in *Nations and Households in Economic Growth: Essays in Honor of Moses Abramovitz*, edited by P.A. David and M.W. Reeder. New York: Academic Press, 89-125.

Eckersley, R. 2005. Is modern Western culture a health hazard? *International Journal of Epidemiology*, 35(2), 252-8.

Emmerson, R.M., Fretz, R.I. and Shaw, L.L. 1995. *Writing Ethnographic Fieldnotes.* Chicago, IL: University of Chicago Press.

Furedi, F. 2004. *Therapy Culture: Cultivating Vulnerability in an Uncertain Age.* London: Routledge.

Geertz, C. 1973. *The Interpretation of Cultures.* New York: Basic Books.

Hanlon, P. and Carlisle, S. 2009. Is Modern Culture bad for our wellbeing? *Global Health Promotion*, 16(4), 27-34.

Hanlon, P. and McCartney, G. 2008. Peak Oil: Will it be public health's greatest challenge? *Public Health*, 122, 647-52.

Hanlon, P., Walsh, D. and Whyte, B. 2006. *Let Glasgow Flourish.* Glasgow: Glasgow Centre for Population Health.

Harrison, P. 1993. *The Third Revolution: Population, Environment and a Sustainable World.* London: Penguin Books.

Hartmut, R. 1998. On Defining the Good Life: Liberal Freedom and Capitalist Necessity. *Constellations*, 5, 201-14.

Horvitz, A.V. and Wakefield, J.C. 2007. *The Loss of Sadness: How Psychiatry Transformed Normal Sorrow into Depressive Disorder.* Oxford: Open University Press.

Hubbert, M.K. 1945. Energy from fossil fuels. *Science*, 109, 103-9.

IPCC (Intergovernmental Panel on Climate Change) 1996. *Second Assessment Report.* New York: Cambridge University Press.

IPCC (Intergovernmental Panel on Climate Change) 2007. *Fourth Assessment Report.* New York: Cambridge University Press.

James, O. 2007. *Affluenza: How to be Successful and Stay Sane.* London: Vermilion.

Lane, R.E. 2000. *The Loss of Happiness in Market Democracies.* London: Yale University Press.

Layard, R. 2006. *Happiness: Lessons from a New Science.* Middlesex: Penguin.

Leyland, A., Dundas, R., McLoone, P. and Boddy, F.A. 2007. *Inequalities in Mortality in Scotland 1981-2001.* Glasgow: MRC Social and Public Health Sciences Unit.

Lury, C. 2003. *Consumer Culture.* Cambridge: Polity Press.

Marks, N., Abdallah, A., Simms, A. and Thompson, S. 2006. *The (un)Happy Planet Index. An Index of Human Wellbeing and Environmental Impact.* London: New Economics Foundation.

McCormick, J. and Leicester, G. 1988. *Three Nations – Social Exclusion in Scotland.* Paper No. 3. Edinburgh: Scottish Council Foundation.

McMichael, A.J., Woodruff, R.E. and Hales, S. 2006. Climate change and human health: Present and future risks. *Lancet*, 367, 859-69.

Nettle, D. 2005. *The Science Behind Your Smile.* Oxford: Oxford University Press.

Offer, A. 2006. *The Challenge of Affluence: Self Control and Wellbeing in the United States and Britain since 1950.* Oxford: Oxford University Press.

Platt, S., Petticrew, M., MacCollam, A., Wilson and Thomas, S. 2005. *Mental Health Improvement: An Appraisal of Scottish Policy.* Edinburgh: Research Unit in Health, Behaviour and Change.

Roberts, B. 2005. *The End of Oil.* London: Bloomsbury.

Ryan, R.M. and Deci, E.L. 2001. On happiness and human potentials: A review of research on hedonic and eudaimonic wellbeing. *Annual Review of Psychology*, 52, 141-66.

Schwartz, B. 2000. Self-determination: The tyranny of freedom. *American Psychologist* 55, 79-88.

Seligman, M.E.P. and Csikszentmihalyi, M. 2000. Positive Psychology: An Introduction. *American Psychologist*, 55, 5-14.

Sennett, R. 2006. *The Culture of the New Capitalism.* New Haven: Yale University Press.

Slater, D. 1997. *Consumer Culture and Modernity.* Cambridge: Polity Press.

Sointu, E. 2005. The rise of an ideal: Tracing changes discourses of wellbeing. *The Sociological Review*, 53, 255-74.

Walsh, D., Benden, B., Jones, R. and Hanlon, P. 2010. It's not 'just deprivation': Why do equally deprived UK cities experience different health outcomes? *Public Health*, 124, 494-5.

Wilkinson, R.G. 1996. *Unhealthy Societies: The Afflictions of Inequality.* London: Routledge.

Williams, S.J. 2000. Reason, emotion and embodiment: Is 'mental' health a contradiction in terms? *Sociology of Health & Illness*, 22, 559-81.

World Health Organization. 2001. *Mental Health: New Understanding, New Hope.* Geneva: World Health Organization.

Chapter 9

Exploring Embodied and Emotional Experiences within the Landscapes of Environmental Volunteering

Stuart Muirhead

This chapter explores the linkages between wellbeing, landscape and environmental volunteering. There will be a focus on understanding natural environments as possible spaces for therapeutic encounter and this will be understood through embodied volunteer experiences. This will incorporate a particularly geographical perspective on understanding relations between landscape and physical and mental wellbeing. The distinctiveness of environmental volunteering is captured by these particular interactions, ones that involve blurred emotional and multi-sensory feelings that are supported by complex motivations and beliefs. These terms are theoretically discussed to begin with and then investigated through empirical research that was conducted with five environmental volunteering groups across Scotland.

The Health Benefits of Nature and Green Space

Growing evidence, especially in the last two decades, has begun to support the view that exposure and access to natural environments can have a wide range of positive impacts on human wellbeing (Burns 1998, Lundberg 1998a, Lundberg 1998b, Pretty et al. 2004, Parr 2005, Townsend 2006, Ulrich and Parsons 1992). For example, Collins and Kearns (2007) outline five ways in which beaches, as areas of natural environment, could enhance wellbeing. These include an enhanced degree of physical or psychological removal from the everyday, an opportunity to be closer to natural environments, providing opportunities for both solitude and social activity and also as a way of shaping collective and social identity. These first four dimensions are drawn from Conradson (2005a) and his work on landscape experiences having a potential therapeutic quality. Collins and Kearns cite the fifth way as being the ability to exercise and carry out physical activity in these spaces. These broad themes are used to frame the environmental volunteering research in the chapter, with a conscious omission of the ability of the volunteering to engender, encourage and promote social capital and social identity. Instead, the focus is on a more introspective view of personal wellbeing, situating itself in physical landscape and within the bodily and emotional responses of the volunteers.

Physical Wellbeing and Embodiment

There have been growing discussions in the literature relating to the physical benefits of exercise in the natural outdoors and the relationship this has with psychological wellbeing (Cooper et al. 1999, Hartig 2008, Pretty et al. 2005). In this chapter, the physicality of tasks and how volunteers experience them will be used to link with aspects of wellbeing. As part of the physical being in the landscape, embodiment is used to further frame this understanding of wellbeing. Embodiment is a concept that assumes the experiences of the individual are shaped by the active and reactive entity that is their body (Parr 2005). Hall (2000) writes about the 'fleshy reality' that affects the physical and mental state of the body and how our experiences of health must be understood through our body as both a biological and social vessel. This bodily experience is central to understanding and interpreting the embodied experience of the volunteer. This initial frame will be used to explain the importance that volunteers place in the physicality of tasks and how they experience them through the very physical presence of their bodies in the landscape and task. This physical wellbeing is closely linked to an emotional wellbeing.

Emotional and Social Wellbeing

Emotional geography is concerned with the association between feelings themselves and the representations and accounts of these feelings that are experienced through the body and within particular spaces. Davidson and Milligan describe the emotional experiences that take place through the body as the most 'immediate and intimately felt geography' (2004: 523). The short term experience of an emotion or emotions can have a cumulative affect on wellbeing. In this sense, wellbeing is the long term wellness of an individual, an underlying sense of being and feeling well or unwell. Therefore, through the physical tasks, an emotional response is experienced and expressed by the volunteer. Using an emotional wellbeing as a second frame for wellbeing, volunteer experiences are explored, illustrating particular aspects of wellbeing that are being experienced by the volunteers.

Environmental Volunteering and Therapeutic Experiences

Townsend (2006) outlines in more detail the mental and social health benefits associated specifically with environmental volunteering. This research emphasizes the importance of not only being in nature and making friends and community connections, but also the significance of reaching a sense of achievement in the work that is done and making a positive contribution to human society. O'Brien et al. (2008) also examine the motivations and personal wellbeing benefits associated with environmental volunteering. Therapeutic landscape literature is used to guide linkages between the volunteering landscape and the other forms of physical and emotional and wellbeing. A number of authors have written about the term 'therapeutic landscape' and the potential benefits of these spaces and places to mental wellbeing (Conradson 2005b, Gesler 1992, Kearns and Moon 2002, Williams 1999). This builds on the idea that certain environments are in themselves beneficial to an individual and their wellbeing. This linkage is not always as straightforward as 'it is not always enough just to 'be' in a place to guarantee a 'therapeutic' outcome: rather there is a need for a skill or artistry in our engagements with place' (Thrift 1999: 310-11).

Clearly the aspects of wellbeing that frame this chapter are interconnected with blurred boundaries between physical and emotional wellbeing. The connections between these two will go towards illustrating the particular characteristics of environmental volunteering that have overall effects on wellbeing. As part of this, both positive and negative aspects of wellbeing are discussed in an attempt to encompass the complex nature of the volunteer experience.

Researching with Environmental Volunteering Groups

The author conducted field research with five environmental volunteering case study groups across Scotland. This was completed as part of an Economic and Social Research Council CASE funded PhD studentship, the Forestry Commission Scotland being the CASE partner. Each of these case study groups facilitated physically active volunteering tasks in both rural and urban green spaces, involving volunteers that were local to an area and also those that travelled to remote locations to volunteer. The experience was captured through an ethnographic strategy that was employed with the case study groups over a period of 14 months. Due to the temporally sporadic and seasonal nature of the organizations, research was undertaken throughout the 14 months, allowing time to build up closer relationships with both organizations and individual volunteers. Participant observation with each of the groups was employed. This often occurred over short task days, but also incorporated more intense residential volunteering weeks in remote locations and included attending committee meetings and other group events. Fifty-two in-depth interviews and four focus groups were carried out across the five case study groups. This ethnographic strategy was designed to build complementary methods that would be able to unravel and reveal the volunteer experiences and begin to unpick the different ways of 'knowing' that are involved in embodiment and emotionality (Davies and Dwyer 2007).

Emotional and Embodied Aspects of Landscape and Task

The term wellbeing in itself implies the placement of an individual in a particular place: 'human existence is only possible through 'being' in the world ... Wellbeing

**Figure 9.1   A volunteer planting Scots Pine trees amongst the heather in
the Scottish Highlands**

also suggests 'being somewhere' (Kearns and Andrews 2010: 309). This idea of
wellbeing involving a physical presence in place is reflected by the significance
volunteers place in where the task is located.

An improvement in wellbeing has been attributed to escaping the pressures of
modern living and gaining a connection to the plants and earth in the form of a
very embodied interaction with the natural surroundings (Edensor 2000). It is this
lasting connection, linking to the volunteering experience, which one volunteer,
Rhona, elaborated on:

> I could really feel as if I could drink the place in today. We got completely
> soaked but I quite enjoyed that. I just couldn't believe that I was actually out
> there in the elements. It's those things that you remember, that was the best part
> of the week so far. I felt like I was completely exposed, feeling the rain, listening
> to it all around me... it was just me out there and that had been the first time I had
> been like that in a long time ... I came up to Scotland when I was younger and I
> felt that freedom again. I hadn't known I'd missed it so much.

Rhona was interviewed during a residential work-week in the Scottish Highlands.
I had observed and recorded her behaviour in my ethnographic diary on the day

she was describing: 'The rest of the volunteer group had been further up the hill at the time and heading back to the mini-bus and Rhona and I were the only two left out on the task. I had watched her for a number of minutes as she stood totally still in the rain, she did not seem to be affected by the pouring rain, or if she was, she seemed to be enjoying it' (Ethnographic field diary). Rhona speaks of the exposure she felt when she was out in the rain, and of the solitude that brought to her. She had previously been nervous about coming on the work-week and was surprised to find how much she had enjoyed working outdoors and highlighted how it brought back feelings and memories of experiences when she was younger, ones that she had missed. It is this interaction between bodily experience in landscape and subsequent mind placement that many of the volunteers explained as valuing most strongly. This would occur, as in Rhona's case, through reflecting on past experiences or a focus on thinking about loved ones. Not only was this through the immersion in landscape and the elements but also through the task that was being carried out.

Tree planting was particularly associated with reflecting on wider life experiences. Some attributed this to the temporal associations with this task. In the majority of cases, the young trees that were being planted by the volunteer groups were between two and three years old and had been cared for in a nursery before being taken to plant on the hillside or park. In these locations, past volunteers had been on the same sites and they were surrounded by previous plantings, at various stages in their lifespan. The volunteers may never be back to that particular site to see the tree grow to maturity, or to see a forest emerge from the heather. The volunteers are therefore playing a small temporal part in the process, a process that was started by people before them, and will be continued by others in the future. This feeling of being part of something bigger comes from working with people of similar beliefs and environmental ethics, but also from being more aware of the extended lifespan of a tree or forest. These thoughts have the potential to be invigorating or humbling, but also painful:

> I've found myself, it's funny I was talking about this today, I've found myself today because we were planting trees, reflecting on the tree planting aspect, the fact that my father passed away and I thought about my father a few times today but I try not to get too hung up on that because it was quite a painful experience at the time when it happened, we were very close so you can dwell on something too much, too long.

In this case, the volunteer was reflecting on the recent death of his father, an event that was fresh in his mind and one that was a very difficult experience for him. He attributes the task of tree planting to contributing towards this being foremost in his thoughts, a time where he is mentally more pensive, through both the nature of the task and the solitude of his surroundings. This emotional focus does not always take the form of thinking about wider life events. Instead many found their volunteering as an opportunity for clearing their minds and thoughts.

Conradson (2007) uses the term 'stillness' to describe areas of (predominantly) natural beauty where people may thus find themselves lost for words, which may reflect an overwhelming sensory or cognitive experience of the present moment. He goes on to describe stillness as a period of calm within an individual, whereby a person becomes more aware of their direct surroundings and less aware of what is occurring in their wider life.

This sensation can happen in any location, Conradson cites a train or plane journey, but he also acknowledges that therapeutic environments and physically distant spaces from day-to-day living may be most conducive to this. Two of the case study groups involved the removal of the volunteers from everyday life, in terms of time, distance and a break from outside communication. This experience of being 'further away' but also being 'surrounded' by the landscape, gave a period of time without distractions and interruptions. However, there was also a sense of this within those volunteers who do not find themselves in as physically distant a location from their wider life. This was attributed by one lady who volunteered in the local 'Friends of' group to the 'ability of a park to be able to change season to season, day to day'. She described being able to see the park enveloped in a thick fog in the morning, then returning later that evening to a clear, moonlit night. Volunteers were therefore given to moments of new discovery and encounter through seeing the park change over the period of a number of hours, but also over the period of a year. The natural processes that occur throughout these times would help in the psychological removal from the every day, even if the physical removal was not as distant when compared to more remote volunteers who travel to the Highlands.

Again, this has linkages with personal and physical notions of wellbeing in the task locations. The relationship is not solely with the landscape, however, it is the interconnectedness between the landscape and the embodied tasks and experiences of the volunteers.

Multi-sensory Encounters

> Derren had led the group over to the wood ants' nest that overlooked the flowing river. I had been here before and the location of the nest between stone, earth, water and an ancient Scots Pine had always struck me. We could hardly hear Derren over the rush of the water so had to move in close to hear what he was saying. He asked Mhairi to run her hand over the nest with the bustling ants on top and then to smell her hand. I had seen this done before but unfortunately she put her hand too close and a few of the ants went in for a bite with their mandibles, quickly followed by a squirt of formic acid. With a yelp of surprise (and perhaps a little pain) the ants were eventually shaken off. When Mhairi did smell her hand she exclaimed that it smells just like salt and vinegar crisps. Derren explained that this was the formic acid that the ants fire as a form of defence... particularly unpleasant if the acid finds an open wound recently made by a bite!

The above account is an excerpt from the ethnographic field diary, one that begins to describe the arrival encounter that volunteers experienced on the first day of a volunteer work-week in the Highlands. Derren, the group leader, is taking a group of 10 volunteers through a small area of ancient Caledonian pine forest. The ants nest provides a very embodied experience for the volunteer. The visual intrigue of the movement of the ants on the nest, the sound of the river, the surprise/shock of the ants' bite and the smell of the formic acid all provide a stimulus for wonder and learning. Combine this with the taste of the seasonal blaeberries and cowberries that cover the forest floor in this area and the sensory experience is complete (of the common senses at least). These multi-sensory experiences were valued by the volunteers in terms of how they physically and emotionally 'know' their volunteering.

There were a number of examples during tasks, where volunteers particularly valued a strong embodied experience where their senses are free to be exposed. One volunteer expressed their resistance to wearing the required safety gear (goggles, hard, hat, gloves and high visibility jacket). He said that he felt he was behind 'some kinda forcefield' when he was conducting the tasks with all of this gear on. He valued being exposed to the elements and felt that this opportunity was being restricted. As well as reducing this sensation, the safety gear could also have more practical drawbacks. One volunteer wanted to feel the 'soil and tree roots in her hands as she planted' and this was restricted by cumbersome rubber gloves. This emotional connection through a physical link or touch was echoed by volunteers throughout the case study groups. Another individual would lie in the deep heather, often without a top on, and watch the weather pass above him. Others would enjoy the touch of a tree, feeling the contoured bark and seeing the life that lives between the bark (insects, lichen, moss, fungi), especially that of the deep 'jigsaw bark' of the Scots Pine. Another volunteer expressed how there was 'nothing better' than reaching into the soil with her hand and pulling out potatoes she had grown herself from the raised vegetable beds that she had planted in an urban meadow. There is a very direct link being made between the bodily sensations that are experienced and how this affects enjoyment of the volunteering. These bodily senses do not just experience space, but also help to structure how that space is experienced. Embodied practices may even have the ability to dominate the volunteering through negative interruptions such as headaches, stomach cramps or mosquito bites (Edensor 2000). There is consequently not always a positive correlation between the physical encounters of environmental volunteering and an enjoyable emotional experience.

Physicality and Wellbeing

The majority of responses to the physicality of the active volunteering tasks were ones of enjoyment, either of the challenge of the exercise or of the benefits of the exercise itself. Volunteers would most often associate the task with an increase in the fitness of their bodies, of doing something that is 'good' for them. The more physically demanding tasks were very different from those such as tree planting or vegetable picking. These included tree felling, deer fence removal, rhododendron eradication and drainage ditch digging. These were often associated with cathartic release. This incorporates the idea that an angry and embodied physical cathartic release can be beneficial to an individual's emotional wellbeing:

> I quite like, I love the planting but I can only do a couple of days of it before I start to get bored. I really enjoy the kinda, sounds a bit dodgy, but I like the destructive side of it. I really enjoy taking out fences and I really enjoy felling.

The challenge of the physicality of the task and the use of positive anger to get the job done was a strong part of this enjoyment. Even the language used by the volunteers in interviews and through the ethnographic diary contained elements of aggression. Words such as 'tearing, 'ripping' and bashing' were used when volunteers described removing invasive plant species or non-native trees. Volunteers would be heard grunting and swearing when tackling a particularly troublesome fence post or sawing through a large tree trunk. These explosive outbursts would quite often be followed by an expression of delight or exhilaration when the task was achieved successfully. The pleasure here is from an instant gratification of physically managing to remove a fence or tree and being able to immediately see the impact this has on the area. This is in comparison with the longer term satisfaction that is experienced by tasks such as tree planting or seed collection. The pleasure taken from the physical nature of the environmental volunteering was not always universal. In some instances a volunteer would view the physicality of the work as being an element that potentially excluded them or made them feel like an outsider:

> I do feel that I can't keep up with everyone else and that I am holding people back. No-one ever says anything but I know I can slow people down. I get frustrated as well because I don't like people feeling obliged to stop and wait for me but if we are walking up a hill and the group has to stay together then they have no choice. If I can avoid that type of task, or ones that involve a lot of walking to get to a location, then I will.

The volunteer, Janet, found it difficult to keep pace with the rest of the group when walking to some of the volunteering task locations. Although the exercise itself may be good for Janet's physical fitness her 'personal disposition' (Kearns and Andrews 2010) to this physical aspect of the group work had negative connotations for her, making her feel distanced from other group members. This idea of personal disposition is one that resonates through many themes of wellbeing. What one individual views as a positive benefit may hold negative impacts for another. The physicality of some tasks may be beyond either the capability or the comfort level of a volunteer's fitness. The volunteers clearly experience the

Figure 9.2   A group of volunteers working together to remove a section of redundant deer fencing

physical embodiment of tasks and relate to them in complex ways, experiencing both positive and negative aspects of wellbeing.

Ethical Underpinnings: Practicality and Spirituality

The motivations that individuals expressed were often centralized in the ethics that supported the volunteer work. There was a sense of stewardship and responsibility for the environment, again providing a temporal link through the past, present and future. This feeling was not only for future generations of people but for the future of the environment, with the individual volunteer situating themselves as an active part of that process. This would be related to achieving a certain goal or contributing towards a specific ecological aim. These practical goals may be related to the upkeep of an urban park, or they may be related to planting a certain amount of trees. The motivation here is to maintain or restore natural environments and the individual contributes to this in a very practical way through their active engagement with environmental volunteering.

In addition to this practical approach, there was also evidence of volunteers linking these goals to a spiritual awareness, both in relation to the environment and to their own physical and mental wellbeing. One volunteer describes this as involving his 'inner and outer self', how he treats his own wellbeing and also how he approaches the environment and others around him. He interconnects these into an overall picture of wellbeing, a way of living his life, linking to ideas of responsibility to himself and others, and also to a stewardship of the environment. Another volunteer speaks about the idea of 'sacredness' and how this has to take into account all aspects of his life. Roberts and Devine (2004) argue that the ability for volunteers to be able to identify and articulate their own satisfactions and motivations is a challenge and it is also argued that within volunteering the experiences are 'emotionally laden, whether or not [they are] also cognitive' (Bondi 2008: 262). The thoughts expressed by these volunteers' ideas hold the same goals as the more practical approach but the difference is in the conscious awareness of nurturing one's own wellbeing and carrying these thoughts through wider life decisions and experiences. This shows a more explicit awareness of what they emotionally gain from their volunteering and also how they express this awareness to themselves and to others.

Conclusion

The field research discussed in this chapter emphasized the importance of 'being' in a certain landscape and experiencing this through the 'vessel' of the body. This is affected by both the place that environmental volunteering takes place and also the type of task that is being conducted. Some tasks are more conducive to solitude and reflection, linking the embodiment of the experience, to the emotional mind placement of the volunteer. This may result in time to reflect on past experiences and memories, be they pleasurable or painful. On the other hand, they may also give the opportunity to have an 'internal state of calm in which a person becomes more aware of their immediate, embodied experience of the world' (Conradson, 2007: 33). These forms of mind placement were most commonly expressed through individual tasks, those that encouraged silence and care. These feelings could occur in both locations that were physically removed from their everyday lives, but also in locations such as city parks and green spaces. The changing daily and seasonal cycles made these possible, providing opportunities for new discovery and psychological removal. In addition, the temporal nature of the task, especially that of tree planting, encouraged a very different relationship with emotionality. Volunteers would link their work to extended timescales and to the influence that their volunteering will have on the landscape for future generations of both humans and non-humans. These emotional connections therefore occur in a number of patterns. They have the ability to connect people and environments over distant times and locations but they can also be fleeting and momentary.

Volunteers value the multi-sensory presence in a landscape, connecting embodied and emotional wellbeing. Importance was put on touch and exposure and in being able to connect physically with soil, plants, life and weather conditions. This physical bond helped structure the volunteering, a frame through which the rest of the volunteering experience may be understood. The physical effort and exertion of many of the active tasks also contributed towards feelings of satisfaction and achievement. Often, this was through an enjoyment of exercise and fitness. However, in contrast to the solitude and reflection experienced in some of the more individual and less physically arduous tasks, these tasks would create a cathartic release. The destructive element and physical achievement, combined with the visual impact of the changes they had made to the landscape, contributed to their feelings of wellbeing. Underpinning these feelings and also the emotional and embodied linkages of wellbeing are the ethics and values that the volunteers hold. The positive impact on wellbeing is related to both practically and spiritually fulfilling their motivations to volunteer. This is expressed through a desire to care for both the natural environment, but also a desire to care for their own internal environment and personal wellbeing.

The volunteering tasks can therefore provide a merging of two very different experiences, those of quiet and calm and those involving physical exertion and even destruction. On first inspection these experiences would appear to be somewhat separated, one being wholly emotional and the other relating to the physical embodiment of the task. However, where these come together is in the underpinning values and ethics that the volunteers hold. These tasks are contributing towards a bigger picture that the volunteers emotionally care about and are bodily engaged in influencing. Active environmental volunteering can therefore be observed as a way in which volunteers express emotional and embodied wellbeing connections with place through the implementation, and fulfillment, of their own core values and beliefs.

References

Bondi, L. 2008. On the relational dynamics of caring: A psychotherapeutic approach to emotional and power dimensions of women's care work. *Gender, Place and Culture*, 15(3), 249-65.

Burns, G.W. 1998. *Nature-Guided Therapy: Brief Integrative Strategies for Health and Well-Being*. New York: Taylor & Francis.

Collins, D.C.A. and Kearns R.A. 2007. Ambiguous landscapes: Sun, risk and recreation on New Zealand beaches, in *Therapeutic Landscapes*, edited by A. Williams. Aldershot: Ashgate, 15-32.

Conradson, D. 2005a. Freedom, space and perspective: Moving encounters with other ecologies, in *Emotional Geographies*, edited by J. Davidson, L. Bondi and M. Smith. Aldershot: Ashgate, 103-16.

Conradson, D. 2005b. Landscape, care and the relational self: Therapeutic encounters in rural England. *Health and Place*, 11, 337-48.

Conradson, D. 2007. The experiential economy of stillness: Places of retreat in contemporary Britain, in *Therapeutic Landscapes*, edited by A. Williams. Aldershot: Ashgate, 33-48.

Cooper, H., Ginn, J. and Arber, S. 1999. *Health-related Behaviour and Attitudes to Older People: A Secondary Analysis of National Datasets.* London: Health Education Authority.

Davidson, J. and Milligan, C. 2004. Embodying emotion sensing space: Introducing emotional geographies. *Social and Cultural Geography*, 5(4), 523-32.

Davies, G. and Dwyer, C. 2007. Qualitative methods: Are you enchanted or are you alienated? *Progress in Human Geography*, 31(2), 257-66.

Edensor, T. 2000. Walking in the British countryside: Reflexivity, embodied practices and ways to escape. *Body and Society*, 6(3-4), 81-106.

Gesler, W. 1992. Therapeutic landscapes: Medical issues in light of the new cultural geography. *Social Science and Medicine*, 34(7), 735-46.

Hall, E. 2000. Blood, brain and bones: Taking the body seriously in the geography of health and impairment. *Area*, 32(1), 21-9.

Hartig, T. 2008. Green space, psychological restoration, and health inequality. *Lancet*, 372(9650), 1614-15.

Kearns, R.A. and Andrews, G.J. 2010. Geographies of Wellbeing, in *The SAGE Handbook of Social Geographies*, edited by S.J. Smith, R. Pain, S.A Marston and J.P. Jones. London: SAGE, 309-28.

Kearns, R.A. and Moon, G. 2002. From medical to health geography. *Progress in Human Geography*, 26(5), 605-25.

Lundberg, A. 1998a. Introduction, in *The Environment and Mental Health*, edited by A. Lundberg. London: Lawrence Erlbaum, 1-4.

Lundberg, A. 1998b. Environmental change and human health, in *The Environment and Mental Health*, edited by A. Lundberg. London: Lawrence Erlbaum, 5-25.

O'Brien, L., Ebdon, M. and Townsend, M. 2008. *Environmental Volunteering: Motivations, Barriers and Benefits.* Edinburgh: Forestry Commission.

Parr, H. 2005. Emotional geographies, in *Introducing Human Geographies*, edited by P. Cloke, P. Crang and M. Goodwin. London: Hodder Arnold, 472-84.

Pretty, J., Griffin, M. and Sellens, M. 2004. Is nature good for you? *Ecos*, 24, 2-9.

Pretty, J., Peacock, J., Sellens, M. and Griffin, M. 2005. The mental and physical health outcomes of green exercise. *International Journal of Environmental Health Research*, 15(5), 319-37.

Roberts, J.M. and Devine, F. 2004. Some everyday experiences of voluntarism: Social capital, pleasure and the contingency of participation, *Social Politics*, 11(2), 280-96.

Thrift, N. 1999. Steps to an ecology of place, in *Human Geography Today*, edited by D. Massey, J. Allen and P. Sarre. Cambridge: Polity Press, 295-321.

Townsend, M. 2006. Feel blue? Touch green! Participation in forest/woodland management as a treatment for depression. *Urban Forestry and Urban Greening*, 5(3), 111-20.

Ulrich, R.S. 1992. Influences of passive experiences with plants on individual wellbeing and health, in *The Role of Horticulture in Human Well-Being and Social Development: A National Symposium*, edited by D. Relf. Portland, OR: Timber Press, 93-105.

Williams, A. 1999. *Therapeutic Landscapes: The Dynamic Between Place and Wellness*. Lanham, MD: University Press of America.

Chapter 10

Place Matters: Aspirations and Experiences of Wellbeing in Northeast Thailand

Rebecca Schaaf

There has been considerable growth of interest in the concept of wellbeing as the desired outcome of development, expressed in the view that 'international development is fundamentally about competing visions of what wellbeing is or should be' (McGregor 2006: 38). This has been combined with contrasting perspectives on how to achieve development, including an increased emphasis on acting collectively coexisting with a continued focus on the individual. The contested nature of the aim of the development process and the debate over the means to achieve this is particularly evident in the context of Thailand. Thailand is experiencing a period of rapid economic, social and political change, with both pressures from international globalising forces and a drive from within to develop and modernize. These tensions are played out in particular in newly emerging peri-urban areas that highlight the diverse visions of, and strategies to achieve, wellbeing.

'Wellbeing is a quality in demand in today's society' (Sointu 2005: 255), yet wellbeing is a complex and contested concept particularly as it is produced in and through specific geographical contexts. As a result, greater appreciation of the relationship between place and wellbeing aspirations and experiences is needed in order to more fully understand the situated nature of wellbeing. This chapter draws on fieldwork conducted in Thailand to explore how national and local characteristics affect wellbeing aspirations, the wellbeing strategies adopted and the ability to satisfy wellbeing goals. In contemporary Thailand there is a national emphasis on the goals of sufficiency and modernization for which group activity is viewed as a key developmental strategy. However, it is at the local community level where tensions between these national development goals and other forces for rapid economic and social change are felt and mediated. As a result, wellbeing experiences and expectations depend on the nature of places and the communities situated within them. This holistic view of the social world recognizes the connection between people and places, and people within places, and the influence each can have on the other. The chapter draws on a case study of a peri-urban village in the Northeast of Thailand in order to explore a variety of community group experiences in detail. The focus here concerns how people value and pursue different wellbeing goals in different ways and make these lifestyle choices within their community. The study connects these wellbeing related choices to activities of participation in local community groups within the peri-urban context.

Wellbeing and Development

There has been increased adoption of the term 'wellbeing' in policy and academic circles, to the point where it is now regularly discussed as the aim of development, the end result of poverty reduction, as well as a process, outcome and condition of being. However, agreement on a definition of the term is complicated by its frequency of use and its adoption within a variety of academic disciplines, guided by diverse epistemological and methodological traditions. In this chapter, wellbeing is understood as having both objective and subjective components, and individual and collective dimensions. Following the approach of the Economic and Social Research Council Wellbeing in Developing Countries (WeD) research group at the University of Bath, wellbeing is viewed as comprising what people have, do, and what they think about what they have and do. The WeD approach views people as social human beings (McGregor 2007) and as acting within the context of a particular place. The incorporation of subjective perceptions of wellbeing, including perceptions of what people have and do, is particularly important as it moves away from prescribing for individuals what ought to make them happy and focuses instead on the achievement of individuals' own kind of wellbeing. The emphasis on the subjective also allows consideration of the meaning given to processes of wellbeing construction and the outcomes achieved. Culture and social organization generate meanings 'through which our relationships are conducted and constrained' (McGregor 2007: 327), and it is these relationships and the meanings imbued within them 'that shape what different people can and cannot do with what they have', and determine the value attached to wellbeing goals, thus impacting on people's wellbeing strategies and outcomes (2007: 327).

Operationalizing a Wellbeing Approach

A key starting point for operationalizing this wellbeing approach was Chen's (1997) work on microfinance services. While Chen does not explicitly discuss wellbeing, she produces a framework of four pathways through which individuals, their enterprises and households experience change. These are: the material pathway, which includes changes in incomes, earning capacity and material resources; the cognitive pathway, including changes in knowledge, skills and awareness; the perceptual pathway, comprising changes in self-esteem, self-confidence and respect; and the relational pathway, through which changes in decision-making roles, bargaining power, participation and mobility are experienced. Viewing Chen's pathways of change as dimensions of wellbeing results in consideration of what people have and do, and their perceptions of those resources and activities within material, relational, perceptual and cognitive areas of life. Chen's framework was also expanded through developing criteria for researching change beyond the individual to encompass groups and the community as a whole. This included studying the characteristics, resources and operation of the groups, along with subjective views of those characteristics and their effects. Using this approach, it is possible to identify which aspects of wellbeing are affected by the community groups, and to what extent.

Development in Thailand

Since the 1970s, Thailand has experienced rapid social, economic and political change, contributing to a contemporary context of dynamism and diversity. This changing environment has resulted in considerable challenges to wellbeing, the construction of which is framed by historical and contemporary structures, processes and events at national, regional and local scales. Contemporary Thai development policy has emerged from a context of mixed fortunes in Thailand and the region as a whole. Remarkable economic growth rates have been accompanied by uneven development, widening inequalities and the persistence of poverty, particularly in the Northeast and the South. In addition, increasing concern was voiced about the effect of the economic-focused growth strategy on families, communities and the environment. The economic boom years of the 1980s and early 1990s culminated in the financial crises that affected the Southeast Asian region in the late 1990s and saw considerable strain placed on the coping mechanisms within society, including increased dependence on the safety net of the village (Baker and Phongpaichit 2005).

The crisis prompted a change of emphasis in the early 2000s and a new vision of Thai development as promoting 'sustainability, health, longevity, learning, empowerment, wellbeing and happiness' (UNDP 2007: 35). The new approach included a greater focus on social support networks and social cohesion, with a move to the local supported following 'evidence that traditional moral economies were resuscitated as the crisis bit and those in greatest need were assisted by their fellow neighbours and fellow villagers' (Parnwell 2002 in Rigg 2003: 112). Rigg (2003) argues that the crisis represented a challenge to Thailand's whole model of development and therefore presented an opportunity to argue for an alternative development vision focusing on self-sufficiency, grass-roots strengthening, and, what Hewison (1997) terms a 'new localism'. In addition, the new development strategy promoted collective action to improve the self-reliance of individuals and communities and placed great importance on unity as a means to prevent the collapse of society. Overall, there has been the promotion of a particular way to be and live, and promotion of particular wellbeing goals, in which unity, discipline and community cohesion are seen as necessary elements of living together as good Thai citizens. There is an assumption that this vision is uniformly shared and that strategies put in place to achieve this, such as the emphasis on group activity and community participation, are universally desired and appropriate.

To achieve these goals, various policies have been implemented including support for group formation and other collective activities, support for saving in groups and increased micro-credit facilities to groups and communities. The purported benefits from such groups are wide-ranging and are discussed across a range of development topics, from participatory rural development and microfinance, to community development, civil society and social capital. The supposed benefits can accrue to both individuals and to collectives (Thorp et al. 2005). They include skills and expertise, confidence, friendship, satisfaction and self-esteem (Kelly and Breinlinger 1996). Groups have the potential to reinforce or construct identities, creating a sense of solidarity and loyalty (Thorp et al. 2005), while Botes and Van Rensburg (2000: 56) argue that 'development in the full sense of the word is not possible without appropriate community participation'.

In Thailand, these collectives included savings groups, rice banks, agricultural groups and cooperative production activities. Increased allocation of government funds at the local level are then distributed through these groups. In particular, these policies were the product of an era of populist policy formulation and implementation by the Thai Rak Thai government. Despite the removal of the Thai Rak Thai party from power, the development approach remains within the King's philosophy of 'sufficiency economy': an approach discussed in the 2007 United National Development Programme (UNDP) report on Thailand. This report highlighted the focus on reducing economic vulnerability while also strengthening and empowering local, particularly poorer, communities. Through its favouring of 'wellbeing over wealth' (UNDP 2007: xvi), the approach clearly recognizes the need to look beyond economic growth and focus instead on enhancing self-development and multi-dimensional change. The sufficiency approach also includes a continued emphasis on cooperative activities, strengthening of communities, and providing financial support through groups. Therefore, this development approach focuses not only on improvements in material aspects of wellbeing through economic growth and increasing incomes, but also on enhancing relational aspects of wellbeing to result in united peaceful communities. In this way, individual quality of life enhances, and is enhanced by processes of community development.

Peri-urban Place: Divided and Dynamic Community Context

The concern in this chapter is how these development goals and wellbeing visions play out in the reality of a particular place: a peri-urban village in Northeast Thailand, or *Isan*. The area suffers from infertile, sandy soils and extensive periods of drought and flooding that combine to give poor agricultural productivity (Parnwell and Arghiros 1996). The region has a history of social marginalization (Phatharathananunth 2006); the dominant view elsewhere in Thailand of *Isan* people is that they are backward, reflecting a perception of Northeast village culture as an obstacle to development, due to its traditional, conservative nature

(Hewison 1997). The region also remains economically disadvantaged, with low per capita incomes and economic power concentrated in Bangkok.[1]

The research location, Ban Lao, is a village with a population of approximately 1000 villagers, situated 17 kilometres to the west of Khon Kaen, the province capital. The contemporary context of the *Isan* region reflects its background as a rapidly changing society and economy, while the village itself has been and continues to be affected by government and NGO interventions attempting to accelerate the development 'catch-up' process. However, this rapid development and modernization focus coexists with the profoundly rural and 'traditional' way of life continued primarily by the older section of the village population. The divisions within the village reflect key tensions existing within contemporary Thailand, including the move away from agriculture, increasing hybridity of livelihoods, generational differences, and the attempt to maintain a sense of 'Thainess'.

Defining Peri-Urban

The existence and form of peri-urban areas in Southeast Asia have attracted increasing research attention in recent decades. McGee (1991 in Hirsch 2009) discusses these *desakota* regions as being formed by the juxtaposition of village and city so they are areas where spatial and social organization and activity take hybrid forms, while Hirsch (2009) describes these as 'peri-urban frontiers' and highlights their role in defining new livelihoods and identities. Rigg (1998) notes the multidimensional blurring of the boundaries between rural and urban, expressed through the physical extension of the metropolitan region, the movement of people and the diversity of their livelihoods and the growing economic interdependence of agriculture and industry. In Thailand, Bangkok is no longer the sole metropolitan region with an extended peri-urban zone as there has been recent recognition of the emergence of peri-urban zones outside regional cities, including Khon Kaen (Glassman and Sneddon 2003 in Hirsch 2009). The physical proximity of peri-urban locations allows greater opportunities to access non-farm income, and therefore makes it possible to maintain 'one foot in the paddy field and the other on the factory floor' (Rigg 1998: 503). However, it is not just economic concerns that drive this hybridity of livelihoods and blurring of rural-urban boundaries. Rigg argues that this diversification rests 'within a wider political and cultural milieu', in which 'aspirations are escalating as expectations rise' (1998: 517), and urban values and social relationships increasingly permeate Thai rural society (Hirsch 2009).

---

1   According to the National Economic and Social Development Board, in 2005, the per capita gross regional product in the Northeastern region was 33,903 baht, compared to a national per capita gross domestic product of 109,696 baht (NESDB 2005). In Khon Kaen province, where the research was undertaken, the per capita gross regional product was 59,978 baht.

The peri-urban nature of the village in this study can be seen in a number of ways. Due to its geographical proximity to Khon Kaen, the village is situated within the rural fringe surrounding the city and therefore benefits from the good links to the facilities and opportunities that Khon Kaen, as the administrative and political centre of the region, provides. The interface of rural and urban activities was clearly evident in the village, with a diversity of occupations, including white collar and low-skilled factory work in the city, subcontracting work from local factories, and an underlying dependence on agriculture. Employment activities in Ban Lao were split between those who worked in the village and those who commuted to employment in the nearby towns and city. For those remaining in the village, rice farming, livestock rearing, and net making were the main occupations, while for those leaving the village, construction labour and factory work were the most common. Strong family and friendship ties were also evident in the relatively small village, as were the stark economic inequalities, expressed particularly through variations in housing size and quality. Overall, the importance of the links to the city was illustrated through the flow of goods, services, and labour, particularly the daily commuting patterns for employment purposes. However, this commuting behaviour existed alongside many characteristics of a village in a much more rural location, highlighting the transitional nature of the peri-urban location.

For Ban Lao, the proximity to Khon Kaen, the improvement in transport links, and the increased ability to afford the transport options, particularly the motorbike, have all increased the role and importance of the city as a market for trade and consumption. This was shown in the frequency of visits to Khon Kaen for a variety of purposes, with these visits particularly undertaken by younger villagers. For this group in particular, the Western-style malls in Khon Kaen and widespread television ownership in Ban Lao have contributed to the increased awareness of the inequalities within and beyond Thailand and a desire for a particular consumer-, and material goods-focused type of development. This type of development is arguably also more focused on the development and wellbeing of the individual rather than the community. These globalizing processes, combined with improved infrastructure and transport options have led to a shrinking of space in terms of distance, with Khon Kaen and other cities and regions in Thailand becoming much more accessible, desirable and achievable, both geographically and in terms of lifestyle aspirations.

These changing aspirations and attitudes reflected wider changes in the individual and collective values promoted and adhered to by villagers and the expectations and levels of satisfaction with the state of the community itself. Older villagers expressed greater satisfaction particularly with infrastructure and service provision due to the amount of change they had experienced over time. In contrast, many of the more negative comments concerning the village's characteristics were expressed by younger respondents, who were also less satisfied with the way of life in the village, particularly in comparison with the larger neighbouring village and nearby Khon Kaen, which many visited regularly for work, shopping or leisure.

It is clear, therefore, that this context provides particular challenges for wellbeing enhancement, due to the tension between the promotion of particular aspirations and wellbeing goals at national level, and the reality of life within a divided and dynamic peri-urban village.

Experiencing Wellbeing through Community Groups

This chapter explores the relationship between achieving this desired development and wellbeing enhancement, and the community groups in the peri-urban context that are encouraged to play a role in this process. This includes uncovering the perceived and real role that such groups play and the variations in their importance for different members of the community. During the fieldwork period, there were eight formal community groups operating within the village, all of which included microfinance facilities. They were all connected in some way to local or national government, through initial funding, ongoing loans, or training and support. However, the participation rates and wellbeing experiences related to the existence of these groups were diverse, and the outcomes for participants were rather different from the government's aim of unity, self-reliance and wellbeing.

Patterns and Dynamics of Participation and Membership

According to group membership lists, 37 per cent of adults in the village participated in a formal community group. However, there were distinct patterns of membership and concentration of particular villagers in groups. According to the field survey data, 55 per cent of members were in two or more groups, indicating an uneven spread of membership throughout the village population. This finding was also supported by the view frequently expressed during interviews in the village that the same people participated both in groups and in general community activities. The evidence suggests therefore that certain sections of the village population were much more active in formal and informal community activities than others. This unequal distribution of participation means that any benefits from group membership were also unevenly distributed within the population, refuting the assumption of universality of membership and outcomes for the community from participatory activities.

Membership also varied according to group function, age, gender and wealth (see Schaaf 2010a, 2010b for more details). There were significantly more female than male members of groups and the largest groups were the two that only provided financial services. Participation was also biased towards certain age as well as gender categories, as particular groups appealed to the interests and needs of men and women of different ages. For men, groups were relevant and important for those in the 16 to 39 age categories, shown by the 74 per cent of male members in the survey drawn from this age range. In contrast, the majority of female members in the survey were over the age of 40 (59 per cent of female members). Membership also varied by wealth, as the poorest villagers had difficulty accessing the groups principally due to a lack of land or key contacts with which to guarantee any loans. This finding was supported by information on house and land ownership: house ownership in Ban Lao was common but 28 per cent of non-members did not own their house compared with only two per cent of group members. There was also a statistical difference between members and non-members in their land availability: for members, the mean area available was 30.57 rai[2] compared with 13.52 rai for non-members, a statistically significant difference (t-test, $p< 0.042$). This finding supports the argument that groups were used disproportionately by a middle section of the village population, as villagers with no or small amounts of land were those who could not either afford or gain access to land or who had sold land for profit during the economic boom. On the other hand, wealthier villagers had access to mainstream banking services therefore had little need for the microfinance facilities offered by the community groups. As a result, the groups were really only relevant and used by a middle section of the village population.

In terms of the groups' operations, there was little real involvement by villagers, and few organized activities or meetings. The groups were run by a small committee or solely by the group leader, with little input from ordinary members. The Thai government's promotion of 'sufficiency economy', self-reliance and unity includes the intention that people will join groups to carry out entrepreneurial activity and join in with others in the community. However, it emerged from interview discussions, that there was limited enthusiasm for active participation in the groups. According to the survey, 66 per cent of members joined groups because they had been asked to do so by friends or relatives and felt obliged to become members. Group members were often unaware of production activities occurring in the group, and expressed little desire for involvement in meetings or other organized events. Overall, there was a lack of real interest in doing things together as a group.

Group Membership and Changing Roles in Thai Society

To explain these patterns of membership and interest, it is important to consider the roles of men and women within the household, society and market in the context of rapid change in Thailand. The dominance of women in the financial groups showed that it is women who often take responsibility for the financial management of the household. This was also shown by the 42.5 per cent of female members who obtained loans to cover household expenses. The number of housewives in the groups, for whom external sources of income may be limited, also illustrates the importance of the potential of groups to provide additional household income and social support. 22 per cent of group members listed one of the main two activities of household members as being a housewife, compared with only 8.3 percent of non-members (University of Bath 2004 and own survey data 2005).

_____

2   Local measurement of land area.

The evidence from Ban Lao also suggests that group functions and operations were filling the gaps left by the changing nature of Thailand's employment opportunities. In contemporary Thailand, there are many employment opportunities for young women in factories which were relatively easy to access due to the peri-urban location, thus reducing their need for groups as spaces for income-earning or social activities and support. There are fewer opportunities for older women who have seen a declining role in agriculture, due to its mechanization and resulting domination by male agricultural workers. Hence the groups in Ban Lao appear to have filled the gap for some older women in providing income-earning opportunities and more importantly perhaps, spaces for social support and affiliation.

It is also important to consider the impact of the peri-urban nature of Ban Lao, particularly as group activity in general was not closely related to the village economy and therefore not very relevant for the majority of the village population. In this village, daily commuting to employment was possible, whether to low-skilled factory employment or white-collar office jobs, hence a greater proportion of villagers did not work in the village and there was less need for the groups to provide employment or employment-related functions.

Diverse Outcomes

The Thai government's promotion of 'sufficiency economy', self-reliance and unity includes the intention that people will join groups to carry out entrepreneurial activity and build community cohesion through joint activities. Groups are also intended to create employment opportunities in villages. However, the wellbeing outcomes of membership did not completely correspond with those intended by the government. The most significant outcome for members was being able to obtain and actually obtaining a loan. Saving money and receiving dividends on shares were other significant outcomes of membership. There were few benefits beyond material outcomes. The few members that were more active in the groups, including leaders and committee members, reported greater gains from membership and women expressed that they experienced a much wider range of benefits from participating than men, including benefits beyond material aspects of wellbeing. However, 'no outcome' was the fourth most common outcome of membership, and it became clear that many members were also dissatisfied with their experience of membership.

In terms of creating employment or supporting economic activity, one group had produced significant effects as it contributed to the continuation of the lifestyle in Ban Lao that depended on sub-contracted work. This work involved the finishing of fishing nets produced by nearby factories and therefore was only possible due to the proximity of the village and the wish of a significant section of the village population to work from home so as to combine sub-contracted work with a continuation of farming activities. The group had introduced safer working practices, access to information and increased the potential for productive negotiations with the government and the factories over rights and pay. It thus supported the structure of employment and the way of life that enabled villagers to work from home, care for children and grandchildren, and continue farming while still earning income.

Wellbeing Strategies and Aspirations: Exploring Non-Participation

An exploration of the reasons for not participating in a group made even clearer the diverse visions of what was required, necessary and useful for achieving wellbeing in the village. In terms of voluntary exclusion, the dominant reasons for not participating were a lack of time and a lack of desire to do so. When asked to explain this further, some non-members adopted an individualistic perspective arguing that membership was open to all, but that they preferred to work alone. They argued that their family came first and that it was not possible to prioritize the family while working in a group. This focus on the family was also mentioned in interviews by group members who commented that people were 'selfish' and 'just think about themselves' in explaining why some did not adopt the community-focused attitude promoted by the government. Another respondent similarly positioned villagers as more focused on their own lives and business, and as a result lacked the time to become involved in community activities. There were also significant generational differences in interest, as younger villagers were much less interested in groups and community activities as a whole.

It can therefore be seen that although non-members recognized that groups are for people who want to be social and do things together, they did not all share this enthusiasm for, or interest in participating. One commented that the benefits that groups offered could be found elsewhere, while another avoided group work as the time demands would result in the neglect of her family, which took precedence over working for others. This illustrates that the government emphasis on collective activities was not relevant to all members of the community, with villagers displaying different capacities and willingness to participate. It also shows different villager perspectives with some focusing on helping and being part of the community, while others were less interested in this aspect of community life and were much more oriented towards family responsibilities and prioritizing the family unit. The variations in participation rates and wellbeing aspirations and outcomes highlight clear differences in priorities as well as diverse strategies for wellbeing enhancement.

These differences in interest and involvement in groups and community activities can be considered another effect of the peri-urban nature of the village. In Ban Lao, certain sections of the population desire the community focus found in more remote, rural areas, while other villagers clearly adopted the more individualistic, household focus characteristic of larger urban areas. For those who were willing and able, participation in these activities is seen as evidence of participation in development activities, and thus a sign of taking action to improve the wellbeing of both the individual and the community. From a government perspective, participation in these promoted activities and groups is also evidence of adherence to, and support for the Thai government. This illustrates the connection between the village and wider structures of governance and notions of national unity. However, the unequal participation rates and lack of interest in groups from sections of the village population demonstrate that the promotion of 'good' Thai citizenship through group endeavours is by no means an accepted ideology for all.

Wellbeing Sacrifices

In addition to finding that not all wellbeing experiences from participation were positive, the research also reveals that there were certain trade-offs between different dimensions of wellbeing made by members, particularly the leaders and committee. These included a trade-off between family time and gaining new knowledge and skills through the group, particularly for those attending meetings outside the village. Certain members also effected a trade-off in material aspects of their wellbeing, as they suffered a decline in income due to putting into place safer working practices learnt through the group. Key members in particular spoke about loss of time, being busy, a lack of time for family, partners complaining and families losing out because of members spending time in groups. Hence, for some villagers, groups were a collective means to achieve individual wellbeing goals that would otherwise be more difficult to achieve without the groups' activities, including feeling part of a group and using small-scale financial services. For others who chose not to participate, group activities would have conflicted with individual activities, including family responsibilities and income earning activities. For these villagers, in the context of relative poverty in Ban Lao, the stable income earning activities took precedence over the groups.

In terms of enhancing unity and community cohesion, the groups were also contributing to divisions and exclusionary processes in the village. For example, the exclusion of the poorest villagers meant that they were unable to access the benefits of the groups, in particular, the government loans that were distributed through the groups. The existence of the groups, the limited membership numbers, and the demands on members' time resulted in the creation of loyalties to groups within the village community. This not only resulted in the creation of factions and divisions, but also threatened loyalties to families and households and as such perversely threatened, not strengthened, community cohesion. This notwithstanding, some did feel that group membership and activities provided an increased sense of belonging to, and pride in, the community.

Conclusion: The Importance of Place

As a response to the economic crisis in the late 1990s, and as part of a stream of populist policies, the Thai government encouraged the formation of groups and created a variety of sources of small-scale credit 'designed to turn farmers into small entrepreneurs' (Baker and Phonpaichit 2005: 259), and create, in doing so, individual wealth, self-reliance and community cohesion and unity. This research indicates that groups can play a role in enhancing aspects of wellbeing through pathways of change, but that the groups are not universally available or accessed, the effects on wellbeing vary and compromises may need to be reached; an awareness of the context is vital in order to understand the motivations, activities and outcomes of group membership. In particular there were distinct variations in membership rates and experiences according to age, gender, and occupation, which can be seen to relate to the changing roles of women and men in Thai society. Variations in participation rates according to measures of wealth were also evident, highlighting processes of exclusion and inclusion related to the social and economic structure of the village; the groups afforded no benefits to the wealthiest villagers and were beyond the economic reach of the very poorest.

The example of Ban Lao highlights the difficulty in achieving the dual goals of material wealth and community cohesion. But it also draws attention to the complexities in defining and prioritising values and goals within a rapidly changing environment at both the national scale through development policy and at the local scale through community and individual responses to policies and schemes. A uniform approach to community development, through promoting group and other collective activity and various microfinance facilities, has not had a uniform effect on individuals in Ban Lao. The research has shown that different people choose whether to join groups for different purposes and experience different outcomes.

The discussion in this chapter has highlighted the role that place plays in guiding these wellbeing strategies and outcomes. Wellbeing experiences and goals are guided by individual aspirations, all of which are underpinned and influenced by place. Places can enable and disable wellbeing goals and also mediate between national visions and strategies and individual and community aspirations and experiences. Place guides what is valued and desired and what can be achieved and how. The study location of a peri-urban setting is particularly characterized by conflicts and tensions between wellbeing visions, strategies and experiences.

The tension between the seemingly conflicting goals of individual wealth and self-reliance and community cohesion can be seen as a direct result of the existence of competing and dynamic development visions in Thailand as well as rapid social and economic change. These visions are expressed through particular national development policies and, in the case of Ban Lao, were mediated at the local level by groups, households and individuals in various ways. These development visions include increased emphasis on self-sufficiency, self-reliance and recognizing and strengthening the village, together with support for Buddhist values. This self-reliant approach has been promoted by the King and has gained in popularity since the financial crisis in the late 1990s and the return to the village and the social safety net of the village economy and society that resulted. The promotion of individual economic development through groups reflects this self-reliant approach, while the groups themselves were promoting the values of honesty, transparency and caring for others.

However, these values are in tension with a more materialist and individual-focused approach to the outcome of development emerging in particular from Bangkok, with a desire for rapid economic growth, high consumption levels and Western lifestyles. This alternative vision was evident also in Ban Lao and Khon Kaen in the ownership of and desire for large pick-up trucks and motorcycles, the increase in mobile phone ownership and the changing way of life from a dependence on agriculture to more urban-based occupations. In Ban Lao, these changing aspirations were particularly seen in the actions and attitudes of the younger respondents and in those with greater levels of education who aspired to own motorbikes, left the village for work, social, and consumption purposes and participated less in community activities. The difference in values and lifestyle was also shown by a significant number of villagers discussing a lack of time to participate due to working outside the village, and prioritizing the family over working with, and for, others in the village. Thus the promotion of group participation is undermined by both the relevance and opportunities of group activities and the changing nature of Thai society towards a greater focus on the individual and the family-unit rather than towards the community for activities and support. It also recognizes the importance of different individual attitudes, aspirations, activities and choices within the context of national policies, strategies, values and goals.

This research highlights a number of key points about the role of place in the construction of wellbeing. First, that the applicability and effectiveness of national development strategies is in part determined by place. The example in this chapter explores the everyday detail of how individuals and collectives are attempting to realize particular wellbeing goals within the context of the national emphasis on the goals of sufficiency and modernization, unity, wealth and wellbeing, yet it is at the local community level where the tensions between these development goals and the rapid nature of changes are felt and mediated especially so in the case of Ban Lao due to its peri-urban characteristics. Secondly, the peri-urban context highlights the conflicts between wellbeing visions, strategies and experiences in Thailand. Ban Lao is an example of a dynamic village in transition, which is divided in many ways. In this research, looking at community group participation has proved a useful way of exploring different visions of a community and community cohesion, and different visions of wellbeing and the means to achieve wellbeing. People value and pursue different wellbeing goals in different ways, and make these lifestyle choices within the contexts of their own community. It also highlights the importance of the local social and economic context within which the policy is being implemented, and the need to recognize the nature of, and value attached to existing relationships and the significance of groups within livelihood strategies. As such, the study of peri-urban Ban Lao illustrates the blurred nature of the boundaries between urban, rural, traditional and modern, and reveals how wellbeing goals are expressed and addressed within the context of a particular place.

References

Baker, C. and Phongpaichit, P. 2005. *A History of Thailand*. Cambridge: Cambridge University Press.

Botes, L. and Van Rensburg, D. 2000. Community participation in development: Nine plagues and twelve commandments. *Community Development Journal*, 35(1), 41-58.

Chen, M.A. 1997. *A Guide for Assessing the Impact of Microenterprise Services at the Individual Level*. Washington DC: Management Systems International.

Hewison, K. 1997. Thailand: Capitalist development and the state, in *The Political Economy of South-East Asia. An Introduction*, edited by G. Rodan, K. Hewison and R. Robison. Melbourne: Oxford University Press, 93-120.

Hirsch, P. 2009. Revisiting frontiers as transitional spaces in Thailand. *The Geographical Journal*, 175(2), 124-32.

Kelly, C. and Breinlinger, S. 1996. *The Social Psychology of Collective Action*. London: Taylor & Francis.

McGregor, J.A. 2006. *WeD Working Paper 20: Researching Wellbeing: From Concepts to Methodology* [Online]. Available from: www.bath.ac.uk/soc-pol/ welldev/research/workingpaperpdf/wed20.pdf [accessed: 25 May 2011].

McGregor, J.A. 2007. Researching wellbeing: From concepts to methodology, in *Wellbeing in Developing Countries: From Theory to Research*, edited by I. Gough and J.A. McGregor. Cambridge: Cambridge University Press, 316-50.

NESDB. 2005. *Thailand Economic Data*. National, Economic and Social Development Board of Thailand.

Parnwell, M.J.G. and Arghiros, D.A. 1996. Introduction: Uneven development in Thailand, in *Uneven Development in Thailand*, edited by M.J.G. Parnwell. Aldershot: Avebury, 1-27.

Phatharathananunth, S. 2006. *Civil Society and Democratization*. Copenhagen: NIAS Press.

Rigg, J. 1998. Rural-urban interactions, agriculture and wealth: A southeast Asian perspective. *Progress in Human Geography*, 22(4), 497-522.

Rigg, J. 2003. *Southeast Asia: The Human Landscape of Modernization and Development*. 2nd Edition. London: Routledge.

Schaaf, R.M. 2010a. Financial efficiency or relational harmony? Microfinance through community groups in northeast Thailand. *Progress in Development Studies*, 10(2), 115-29.

Schaaf, R.M. 2010b. Do groups matter? Using a wellbeing framework to understand collective activities in northeast Thailand. *Oxford Development Studies*, 38(2), 241-57.

Sointu, E. 2005. The rise of an ideal: Tracing changing discourses of wellbeing. *Sociological Review*, 53(2), 255-75.

Thorp, R., Stewart, F. and Heyer, J. 2005. When and how far is group formation a route out of chronic poverty? *World Development*, 33(6), 907-20.

UNDP. 2007. *Thailand Human Development Report: Sufficiency Economy and Human Development*. Bangkok: UNDP.

University of Bath. 2004. *Resources and Needs Questionnaire (RANQ)*. Wellbeing in Developing Countries ESRC Research Group (WeD), University of Bath.

Chapter 11

Wellbeing in El Alto, Bolivia

Melania Calestani

In recent decades there has been an increasing worldwide interest in developing universal definitions of wellbeing in academic, governmental and non-governmental circles. Research projects and books on wellbeing have multiplied in the last years; governments have also focused their policies on this concept, often proposing a universal understanding and definition of the latter. There are different examples of this 'wellbeing' policy shift: for instance, in the Kingdom of Bhutan with its gross national happiness indicator and, recently, in Evo Morales' Bolivia.

During the last decade Bolivia has witnessed important political changes. A new constitution was approved in 2009 – Constitución Política del Estado – which placed importance on political discourse about indigenous and national wellbeing, defined as 'Suma Qamaña'. This concept was proposed by some Aymara intellectuals and could be translated as '*buen convivir*' – living well together (Albó 2009). This co-existence implies harmonious relationships not only between human beings but also between human beings and nature. Exploring the concept of *Suma Qamaña* is a highly relevant topic for academic engagement at this moment, both as one of the priorities of the Morales government and as an alternative image of indigeneity. In particular, such exploration needs to critically examine the limitations of this political discourse by distinguishing it from the original meanings of the term as rooted in people's everyday practices, beliefs and culture.

This chapter describes and interrogates the concept of 'the good life' as practiced and articulated by Aymara informants in the city of El Alto. The data are drawn from ethnographic research conducted whilst living in two different neighbourhoods of this city, Amachuma and Senkata, which enabled the inclusion of a variety of perspectives. In particular, research in these two different neighbourhoods affords a perspective that qualifies the term 'urban'. In this chapter, I make the case that El Alto is an excellent field site in which to investigate the complexity of issues that emerge when considering the relationships of wellbeing and place. In Bolivia, the Western and Eastern parts of the country diverge from one another, just as the city does from the countryside and the mountains from the tropical regions. These places differ not only in terms of their ecological space, but also in terms of their social and cultural landscapes, the spaces of collective meanings.

My interest in the topic was stimulated by the increasing interest amongst local intellectuals in investigating wellbeing from an 'indigenous' point of view. In the last decade, there has been an attempt on their part to verbalize a 'collective' concept for the indigenous 'community'. Interestingly, their view depicts 'indigenous' people as a version of Rousseau's 'noble savage'; the countryside is idealized in opposition to the city as the place where 'the original community' and harmonious social relations are destroyed.

Despite the support of my informants to this idealized depiction, my ethnographic work in the city of El Alto in 2004 shows how there are contradictions that emerge from everyday life which influence my informants' practices. The concept of 'the good life' as theorized by local intellectuals is an ideal. My informants know what *Suma Qamaña* means, but usually do not use the term in everyday life. I heard local leaders using it in political speeches to invoke the unity of all the residents. *Suma Qamaña* is identified by my informants as '*bienestar*' -wellbeing or 'the good life' at the community level – but also as '*permanecer*' –to stay and to remain. Many conveyed to me that this ideal concept is difficult to achieve, involving political and economic aspects, in the city as well as in the countryside.

On the other hand, local intellectuals emphasize the existence of an opposition between the rural and urban space, claiming that 'community' does not exist in the city, because of the decline of historical forms of cooperative labour. However, little attention is given by local intellectuals to other communal institutions in the urban context that have replaced or at least compensated for rural organizations. The *Junta Vecinal* and *Junta Escolar* are fundamental urban institutions; the former is the local residents' committee, while the latter is the organisational committee for a school and usually, both these institutions fight for the 'self-construction' of the area (Lazar 2002, Cottle and Ruiz 2000, Urton 1992).

The aim of the chapter is therefore to shed light on how place is symbolically constructed in relation with wellbeing and how 'the good life' assumes different forms even within the same city. The next section provides a brief general background of the Bolivian highlands. Subsequent sections develop the argument through the case study material on El Alto, the neighbourhoods of Senkata and Amachuma.

Living in the Highlands

Unfortunately, people who don't know Bolivia very much think that we are all just Indian people from the west side of the country, it's La Paz all the image that we reflect, is that poor people and very short people and Indian people ... I'm from the other side of the country, the east side and it's not cold, it's very hot and we are tall and we are white people and we know English so all that misconception that Bolivia is only an 'Andean' country, it's wrong, Bolivia has a lot to offer and that's my job as an ambassador of my country to let people know how much diversity we have. (Gabriela Oviedo Sarrete, Miss Bolivia, Declaration at Miss Universe Contest in Quito, Ecuador, May 2004)

Some might argue that a beauty pageant contestant is not a good representative of public opinion. However, Gabriela Oviedo made explicit the tensions existing between different regions, and expressed the cultural and geographical diversity of her country. Bolivia is divided into nine administrative departments. The tropical lowlands in the east are in stark physical contrast to the Western Andean plateau and peaks or highlands and in the centre there are green valleys. These contrasts in the landscape are echoed by cultural differences: there are 34 ethnic groups that are officially recognized. Quechua and Aymara are the larger language groups, although it is questionable how far linguistic markers represent in themselves distinctive and identifiable ethnic affiliations.

Gabriela Oviedo's words suggest that the Bolivian national body is characterized by discrimination and misunderstanding between different regions which are also associated with different social classes and ethnic groups or in her words, 'poor' and 'indigenous'. Miss Bolivia 2004 is from the city of Santa Cruz, which has had an economic boom in the last two decades. It is in the richest region of Bolivia and in June 2005, it attracted media attention because of racist attacks towards indigenous migrants from the highlands who had moved to the lowlands to work in factories. The use of the term *indios* – which carries derogatory connotations – also implies that the image of La Paz, and in general of the Western part of the country and the Andes, is associated with an idea of backwardness and racial inferiority, and that these are discursively linked.

My story is an 'Andean tale' that took place at an altitude of 4,000 metres on a dry plateau surrounded by snow-peaked mountains, a harsh region that farmers have cultivated for millennia and where an imposing new city is growing today. This is El Alto, populated largely by Aymara migrants from the countryside and thus the 'most Aymara' city of Bolivia, and the most indigenous city of Latin America. Until 1985, it was part of La Paz, but since then it has been an independent city. With a population of almost one million inhabitants, it is a hectic hub with colourful markets where you can buy used and new goods from China, the United States and other South American countries.

At the time of my fieldwork in 2004, El Alto was often associated with the demonstrations and roadblocks that had taken place during the last 14 years in Bolivia. Its geographical position makes it a strategic location; El Alto surrounds La Paz, which is positioned 400 metres lower down in a canyon. It is therefore the only accessible route for goods and supplies in and out of the capital and blocking the main roads that link the two cities interrupts the economic flows into the capital. However, El Alto's reputation of combative endurance is not only favoured by its geographical position, but also by the social characteristics and capacity of its inhabitants to organize.

The first time I visited Bolivia in 2002, I was fascinated by the Bolivian highlands. The Andean plateau with its overshadowing mountains is a place where each element is believed to be alive and to have a spirit. The snow peaks have names and personalities; the *Achachilas* – God mountains – are believed to love and hate as human beings do. They are considered to be active members of the community, holding positive and negative feelings as well as desires and appetites. The earth itself – *Pachamama* – is alive and has to be continuously fed, so that in turn it can feed human beings.

Life on the plateau is regulated by the cycle of agricultural production, cattle breeding, sheep farming, and the exchange of products in the market. Fiestas are numerous throughout the year. They indicate the passing of time and the celebrations for the harvest. The landscape of the flat plateau is suddenly interrupted by the red brick houses of the city of El Alto. In the last 25 years, El Alto has become the most attractive destination for rural migrants. Many have left the countryside in search of a better life, because agricultural production was too scarce to support their families or because they felt attracted towards this large anthill of buildings and people. For such migrants, El Alto is associated with better work and education opportunities. This is especially true for young people, who make up the majority of the population. El Alto can indeed be seen as a city characterized by a strong presence of youth and a vibrant youth culture. It has also assumed an image as the 'city of the future' or '*ciudad esperanza*' (city of hope) as many inhabitants and rural migrants like to define it. Nevertheless, this view is not shared by everyone. It is quite common to hear people in La Paz say that El Alto is a poor shantytown that should be avoided, picturing all *alteños* as 'poor indigenous criminals'.

A Brief History of Urban Space in El Alto: Migration, Marginality and Identity

El Alto's history mirrors and illustrates world trends. In 1800, only 1.7 per cent of the 900 million people in the world lived in cities of 20,000 or over (Hauser and Schnore 1966). In the last 150 years, this has changed radically. In the global south many cities are growing at rates so high that they double their populations every 10 to 15 years. This constitutes one of the most important migrations in human history, creating 'overurbanization' or 'hyperurbanization' (Perlman 1976). Comparative poverty in rural areas is a crucial factor, supporting the 'push' theory of migration that focuses on the difficult conditions in the countryside and the changes in agricultural production, such as lack of good land or a low productivity (Gordonava 2004, Cortes 2004).

However, when the *campesinos* (peasants) migrate from their *pueblos* (villages) to El Alto, they do so for several reasons. Although there are differences between the various provinces, *campesinos* generally migrate because they hope to acquire an economic stability that does not exist in the countryside, as many of them told me. But they also migrate because they want to join other members of their family who previously migrated or for access to a better education or social status – all examples of 'pull' factors (Albó et al. 1981). These educational and social aims are an important element in attraction towards the city and El Alto thus becomes '*la ciudad prometida*'– 'the promised city' (Sandoval 1985), where Aymara migrants can improve their quality of life and can give shape to their aspirations.

Initial problems faced by the migrants relate to difficulties in finding employment and accommodation, especially for those without family members who have migrated to the city before them (Albó et al. 1981: 119). Young women between 17- and 25-years-old usually work as domestic servants in the homes of families in La Paz. On the other hand, older women are most likely to be involved in market activities, as *comideras* (preparing hot food and drinks in the street) or as street market vendors.

Migrants are seen as being between two worlds; this peculiar position has been classified as '*el mundo cholo*', where *cholo* means an Indian who has come to live in the city and is somewhere between being Indian and being *mestizo*. It is both a racial and a social category, signified by indigenous physical features combined with particular clothes and economic activities in commerce (Lazar 2002, Harris 1995). Usually, the term *cholo* is counter-posed to the term *mestizo*. They both represent intermediary categories, but while the former indicates closeness to the indigenous set of values, working activities and social position, the latter underlines the adoption of a set of non-indigenous values and working activities, together with a higher social status. Women working in the markets are always classified as *cholas* or *cholitas*, and wear a layered gathered skirt (*pollera*), a shawl, specific pumps for their feet and a bowler hat. Such women refer to themselves as '*mujeres de pollera*' in contrast to '*mujeres de vestido*', that is women in Western clothes. However, the boundaries of social categories are impossible to define in an absolute way. Ethnic ascription must be viewed as situational, as a social process, depending in part on circumstance. Thus, ethnic terms change their meaning according to speakers and context and vary in different parts of the Andes (Weismantel 2001). As Xavier Albó (1998) argues, it would be more accurate to describe the people of El Alto as urban Aymara, a term that I prefer. However, sometimes ethnic terms cannot be avoided, especially when adopted by research informants. Their use shows not only a subjective dimension that cannot be generalized but also more complex and context-dependent elements that flavour ethnic terms with different nuances.

'Migration emerges as a process of self-empowerment, or loss of power, and places become the symbol of embodied phenomenology of space and time' (Napolitano 2002: 4). El Alto is one of those places. Whilst on the one hand it is pictured as '*la ciudad prometida*' – 'the promised city' – by its inhabitants, on the other hand it is defined as 'ciudad en emergencia' – 'emergency city' (Antezana 1993) by the people of La Paz. According to a document produced by the municipality of El Alto in 1997, the major needs of the population relate to the lack of basic services, such as basic sanitation (78 per cent without), appropriate housing (73 per cent), health (68 per cent) and education (64 per cent) (Lazar 2002). El Alto is also a city characterized by high levels of domestic violence and gang participation (Cottle and Ruíz 1993, Revello Quiroga 1996, Pérez de Castaños 1990).

The depiction of El Alto in terms of violence and need has been correlated with issues of ethnicity as well as social and cultural factors; the majority of the migrants are Aymara (Albó 1998) and are employed in the informal sector, which has expanded since 1984 because of hyperinflation (Antezana 1993). Therefore, the image of El Alto as migrant and Aymara sits alongside the informality of the economy, not only implying that these problems are the inhabitants' own fault, but also highlighting structural issues of discrimination and internal colonialism (Cottle and Ruiz 2000, Lazar 2002). By 1995, more than 40 per cent of the economically active population was working eight or more hours a day, but earning less than the minimum wage. Among those employed, women's pay was, on average, 50 per cent less than men's (DFID 2002).

Nonetheless, the informal sector represents a fundamental, if not the only, source of employment for the inhabitants of El Alto. In an informal economy, it is quite difficult to calculate the earnings of workers; in the case of my informants, most of them said that they were earning less than the minimum wage. However, these perceptions and comments may be connected with cultural factors, such as the fear of creating jealousy among neighbours and consequent conflict with the rest of the community. Therefore, while this indicator may be helpful for general discussions on the topic of poverty, it is difficult to calculate, not only due to the conditions of the informal economy but also for cultural reasons which affect reporting of earnings. Moreover, earnings may also fluctuate depending on the various moments and work opportunities.

Prospering in Bolivia's informal sector has always required a high degree of flexibility due to macro-constraints such as changing governmental regimes and periodic monetary crises, as well as micro-constraints such as illness and the consequences of laws and theft. Sofia Velasquez, a Bolivian market vendor and the main character of the books by Hans and Judith-Maria Buechler developed efficient and powerful strategies. For instance, she took advantage of international price differentials and her contacts enabled her to obtain staples in periods of scarcity. Moreover, she substituted the products she sold with other items in order to change unfavourable conditions. Nevertheless, in spite of her talents as a trader, she often felt vulnerable. The desperation with which she reacted to the theft of her items was indicative of the importance and difficulties of maintaining working capital (Buechler and Buechler 1996).

El Alto is a city where many foreign and national elements are fused and locally reconstructed. Understanding *alteño* identity means engaging with stories not only of migration, displacement and marginality, but also of expectation, hopes, social advancement and return. The choice of my two field sites, Senkata and Amachuma, provide important examples of the intricacy and complexity of all these issues.

Amachuma is essentially a suburb of El Alto and people commute daily into the city to work, returning to the zone at night. Although nowadays they are also part of El Alto municipality, Amachuma residents like to define themselves as 'peasants' and see their area as ' a village in the countryside'. It is interesting to notice how they identify their values as 'rural' and distinct from the urban space, from neighbourhoods such as nearby Senkata. Yet, at some points Amachuma and Senkata become the same world, a world in the process of forging a new Alteño identity. The construction of Amachuma's 'rural nature' by its inhabitants, despite their commuting activity and being part of El Alto municipality, is significant and demonstrates the potential for intellectual ideologies, such as that of Aymara intellectuals, to impact on ordinary people. The following sections describe Senkata and Amachuma further in order to convey an impression of how they look and everyday life, an impression which is fundamental to unfolding the complexity of issues emerging when engaging with wellbeing and place.

A Neighbourhood in the City: 25 de Julio, Senkata

Senkata, which in Quechua means nose, is a neighbourhood in the south of the city of El Alto. It belongs to District 8 and has approximately 2600 households, and 18,000 inhabitants, or *vecinos* (a Spanish word that means resident and neighbour). There are different residential areas in Senkata. I lived in 25 de Julio, where there are approximately 400 households and 2,800 inhabitants. Senkata is divided in two by the highway to Oruro and it is bordered to the north by El Kenko zone, to the south and west by Provincia Ingavi, and to the east by Achocalla municipality. The focal point of 25 de Julio is the football field and the market place of the Extranca. The main public buildings are organized around the Extranca: the church; the local residents' committee centre; the Colegio 25 de Julio; and the main local shops. These are the main points of daily interaction between different residents. The Extranca gets very crowded and busy on Thursdays and Sundays, when the main market takes place. On the highway to Oruro, at the level of the Extranca, there are 28 permanent kiosks grouped into a sellers' organization.

According to some informants, Senkata land belonged to *Achocalla hacienda* (landed estate). After the Agrarian Reform in 1953, the land was divided among the peasants who were attached to the landowner. In 1979, the first migrants began their settlements in what is today Senkata 79. The neighbourhood of 25 de Julio was formed later, at the beginning of the 1980s. Senkata residents are first and second generation migrants: the majority of them (39 per cent) have come from other areas of El Alto and La Paz, and 18 per cent come from the nearby Aroma province. Others come from different provinces such as Pacajes, Loayza, Oruro, Ingavi and Inquisivi. Senkata residents work mostly in the informal sector. The most common professions for men are those of builder or bus driver.

Senkata roads are unpaved and very dusty, apart from the highway that goes to Oruro. Houses have access to electricity and water, but still lack a sewage system. The houses are predominantly made of adobe bricks with corrugated iron roofs. They consist of up to four one-room buildings arranged around a courtyard, where there is a tap that provides for the water consumption of the entire household. Everyday life begins quite early as people get up around 5 or 6 am. Streets become busy between 7 and 9 am, when children go to school and their parents to work. Dogs sit outside their houses during the day, menacing any stranger passing by. Some of them can be extremely aggressive, and extra care has to be taken when taking side roads that one does not know well. Everybody is scared of dogs in El Alto. Children learn very quickly how to defend themselves and, with particularly aggressive animals, to skilfully throw stones. Children get together at school either in the morning or in the afternoon and play together in the streets. Adults interact daily in shops and at the market place and get together occasionally when the local residents' committee or the *Junta Escolar* (parents' association) meet. Usually, adults also meet up in the weekly religious services of the different Churches (such as *Asamblea de Dios de Bolivia, Iglesia Adventista del 7 Día, Nazareño Church, Iglesia Evangélica Presbiteriana, Iglesia Católica San Francisco de Assisi*).

Amachuma: 'Rural' El Alto

Amachuma was part of Achocalla municipality until August 2005 when it became the 10th district of El Alto or Distrito Rural Número 10. It is on the road that leads to Oruro, 20 minutes by bus from Ventilla. While in the past it was a *pueblo* or village, nowadays it is officially part of the city. Surrounded by fields, Amachuma has a 'rural soul'; this is evident through its inhabitants' daily life as well as through the landscape. The houses are scattered in the countryside, not neatly arranged as in Senkata.

The example of Amachuma also shows how the city is expanding towards the surrounding countryside and illustrates the diversity of *alteño* contexts, and the many nuances that form alteño identity which cannot be treated as a single identity, but as many and dependent on the area and its local history. Central El Alto, which the people of Amachuma call 'La Paz', looks like a mirage in the desert but its presence is always dominant. To the south of the suburb, there are other rural communities linked to Amachuma by a valley and overlooked by the Andean chain including the snow peak of Mount Illimani.

There are 794 inhabitants living in the community and their mother tongue is Aymara (Instituto Nacional de Estadística 2003). The majority of the population is children (0- and 9-years-old) and young people (10- and 19-years-old). However, when I refer to young people I include people in their late twenties who are unmarried and therefore still seen as young, especially in Senkata, where marriage normally takes place when people are in their twenties. In Amachuma, young people tend to cohabit with their partners or get married in their teen years, becoming adults earlier. There are relatively few older people in Amachuma.

The main occupation of Amachuma inhabitants is agriculture. Livestock are also another important resource, including cows, sheep, chicken and a few llamas and pigs. Women usually sell milk and other agricultural products, such as potatoes and alfalfa, in the markets of El Alto. Men usually work as minibus drivers. Both men and women also earn money from handicraft; women make wool gloves and *tejidos* (woven cloth) and men make pottery (these products are sold in the market). The majority of the houses are made of *adobe* and *tapial* (mud) with an earth floor pavement in most cases. Only in a very few cases is the floor made from cement or paved.

Amachuma inhabitants speak about their *entorno* (environment) as something that is very important for them and feel they are in contact with nature and with the rest of the community. Solidarity among the members of the community is seen as very important and manifest through the practice of Aymara cooperative work, such as *ayni* and *mink'a*. *Ayni* is a reciprocal and obligatory service, for example beer bought for weddings or other festivities (Lazar 2002). "The ayni principle is as follows: I request and receive a specific service from my kinsman or neighbour, on the understanding that on a subsequent, exactly similar occasion, I will render him/her exactly the same service" (Harris 2000: vi). However, in recent years, *ayni* as service has lost its importance because of lack of time, now devoted to activities in other neighbourhoods of El Alto. Yet, *ayni* as gift in religious festivities is still practised. On the other hand, *mink'a*, which describes communal work that is not remunerated and takes place in public spaces, such as paving the main square, is still quite common.

Amachuma men come together in occasional *mink'a*. Relations with kin men are also expanded when they meet in local shops to drink a glass of beer, while women meet in private houses to knit and weave together. Sometimes, the officers of the Peasant Trade Union, which is the most important local administrative unit (see the following section for more details) invite the community to regular as well as occasional special meetings to talk about political activities, such as the protests of October, or other issues concerning communal life. The 'community' also meets on the days of fiestas, religious festivities, and at the occasional meetings organized by the Junta Escolar. In contrast with Senkata, weekly religious services are not attended by Amachuma inhabitants. During my fieldwork, the Catholic Church was always closed, even during the *fiesta* –religious celebrations, and the Pentecostal weekly events at the Baptist Church were rarely attended. On the other hand, Amachuma inhabitants often organize rituals and offerings for the God mountains – Achachilas – and the Mother Earth – Pachamama.

Rural and Urban El Alto

The peasant community with its territory and rules is a colonial creation with the aim of organizing and managing natural resources. Some of these communities were created on the basis of the original *ayllus* (pre-Inca and Inca administrative units), also known as 'original communities' (*comunidades originarias*). In highland Bolivia the term *ayllu* refers to land-holding units (Abercombie 1998, Harris 1987) whose land could sometimes be dispersed over a wide area and occupied jointly with other *ayllus*. Other communities were the result of the transfer of people to territories away from their original location. At the end of the 19th century, the shift to the term *comunidad* (community) occurred with the expansion of the hacienda. After the Agrarian reform in 1953, the notion emerged of the communal property of a territory with defined borders, and the beginning of conflicts between communities and haciendas (land estate) over the right to use resources (Rengifo Vasquez 1998).

Being a member of the *ayllu* means that people have to perform certain rituals and practices (*costumbres* as Amachuma peasants call them). These practices are communal and involve participating in the local *fiesta*, in the *ayni* (reciprocal work and gift in a fiesta), in the *mink'a* (communal work), and in rituals aimed at the communication with supernatural forces.

After the Agrarian Reform in 1953, *ayllu* structures were replaced by the Trade Union, especially in areas dominated by haciendas. In Amachuma, peasants own their own plots of land and they also have access to communal land if they are part of the Trade Union. There is rotating cultivation and the community decides which crops will be grown on communal fields allocated to individual households each year. Communal land can be cultivated, used for grazing or for other ritual and social activities such as the *apachita* (sacred place where supernatural forces are worshipped). If a household wants to sell land to someone outside of the community, the head of the family has to contact the Trade Union and ask permission for the sale. However, there is a tendency not to sell to people outside the kinship network. Even when people decide to migrate to other areas, they prefer to keep their land and to return there once in a while or to leave other members of the family to take care of the property.

The Trade Union decides the rights and obligations of its members, such as participation in communal work, rotation of responsibilities (*cargos*) and payment of fees, as well as the rituals related with agricultural production. The Trade Union in Amachuma intervenes in infrastructural works, such as the maintenance of roads, access to drinking water and the maintenance of the school building but does not actually take part in agricultural production, which is carried out by individual households.

By contrast, in Senkata, it is the *Junta Vecinal* or residents' Committee which plays the role of the local administrative unit and there are important differences between the two zones. In fact, in Senkata there is no use of common land with the exception of the market place and no agricultural production, since it is too densely populated. Women who work in the market place have to pay fees for their *puestos* (stalls) to the *Junta Vecinal*. The majority of my informants own the house where they live and the land on which it is built, their *lote*, and do not have to ask for the Junta Vecinal's permission when they wish to sell. The few people who do not have enough money to buy a house usually work as caretakers of the building in which they live. In this way they do not have to pay rent and owners of the houses, who are scared to leave second houses empty because of thievery, are ensured some security. Owners may also leave young members of the household to take care of the house rather than contracting someone who is unrelated. Similarly, some young people sleep in the local shops that belong to their household to protect the family's resources from theft.

Conclusion

This chapter has presented a description of the city of El Alto and outlined some of the issues emerging when dealing with wellbeing and place. As mentioned at the beginning of this chapter, exploring the concept of *Suma Qamaña* is a highly relevant topic for academic engagement at this historical moment, when Aymara intellectuals are defining indigeneity in terms of 'rurality' or as 'living well in the countryside' – in harmony with the other members of the community and with the natural world. Yet, my analysis critically examines the limitations of their political discourse, as well as of that of Morales government.

The differences among the various areas and neighbourhoods of El Alto enable an exploration of the meanings that my informants attach to the terms 'urban' and 'rural', which inform the differences and relations between Senkata and Amachuma, in spite of their administrative and geographical proximity. The 'rural–urban' framework shows how these 'rural-urban' constructions and classifications assume a very important role for my informants and are often invoked in narratives of modernity/backwardness as well as of morality/ immorality.

This is evident in both field sites. In Senkata, people emphasize the theme of modernity versus backwardness, while in Amachuma they talk more of morality versus immorality. Therefore, Senkata residents picture their neighbourhood 'as modern' in opposition to the 'backward countryside'. On the other hand, Amachuma inhabitants reproduce an account of their Distrito Rural as 'morally nuanced' in opposition to the 'immoral urban space'. The Amachuma residents are 'peasants-commuters', highly dependent on the city and part of it, but also particularly proud to affirm their rural roots and the consequent adoption of certain values in opposition to Senkata and the urban space in general. This illustrates clearly the discursively constructed character of urban-rural distinctions which in spite of the proximity of Amachuma and Senkata, are mutually constituted in relation to one another and in relation to questions of *Alteño* identity, and to ideologies of *Aymara*-ness.

In both field sites, there are different layers of identity that emerge from people's narratives and many contradictions between what people say and do. Thus, a straightforward dichotomy between urban and rural represents other oppositions: past/present, indigenous/non-indigenous or modern/backward. Despite the fact that these are often invoked in my informants' narratives, this chapter has shown that everyday life in El Alto is more nuanced and that the borders between opposites are often highly blurred, with the result that *alteños* are not a homogenous group. Finally, it is increasingly evident that wellbeing requires and is shaped by engagement with culture-specific ways of how to be well. This line of research may eventually serve to illuminate the multiple and fundamental ways in which wellbeing interacts with place, culture and politics.

References

Abercombie, T.A. 1998. *Pathways of Memory and Power: Ethnography and History Among an Andean People*. Madison, WI: University of Wisconsin Press.

Albó, X., Greaves T. and Sandoval G. 1981. *Chukiyawu: La cara Aymara de La Paz*. I. El Paso a la ciudad. La Paz: CIPCA.

Albó, X. 1998. *La Paz es tambien Chuquiyawa. La Paz 450 Anos (1548-1998) Tomo I*, Mesa Figueroa, J., Gisbert, T. and Mesa Gisbert, C., La Paz: CEBIAE, 21-30.

Albó, X. 2009. Suma Qamaña = el buen vivir. *OBETS. Revista de Ciencias Sociales*. N. 4, Universidad de Alicante. Instituto Universitario de Desarrollo Social y Paz.

Antezana, M. 1993. *El Alto desde El Alto II. Ciudad en Emergencia*, La Paz: Unitas.

Buechler, H. and Buechler, J-M. 1996. *The World of Sofia Velasquez: The Autobiography of a Bolivian Market Vendor*. New York: Columbia University Press.

Cortes, G. 2004. Una ruralidad de la ausencia. Dinámica migratorias internacionales en los valles interandinos de Bolivia en un contexto de crisis, in *Migraciones Transnacionales. Visiones de Norte y Sudamérica*, edited by A.H. Gordonava, La Paz: PIEB, 167-99.

Cottle P. and Ruiz C.B. 2000. *Violencias Encubiertas en Bolivia*. Centro de Documentación e Información Bolivia. La Paz: CIPCA – ARUWIYIRI.

Cottle, P. and Ruíz, C.B. 1993. *Violencias Encubiertas en Bolivia*. Centro de Documentación e Información Bolivia, La Paz: CIPCA – ARUWIYIRI.

DfID. 2002. *Bolivia – Country Strategy Paper 2002*. London: DfID.

Gordonava, A.H. 2004. *Migraciones Transnacionales. Visiones de Norte y Sudamérica*. La Paz: PIEB.

Harris, O. 1987. Dinero y Fertilidad. *Economia Etnica*. La Paz: HISBOL.

Harris, O. 1995. Ethnic Identity and Market Relations; Indians and Mestizos in the Andes, in *Ethnicity, Market and Migration in the Andes*, edited by B. Larson, O. Harris and E. Tandeter. Durham, NC: Duke University Press, 351-90.

Harris, O. 2000. *To Make the Earth Bear Fruit: Essays on Fertility, Work and Gender in Highland Bolivia*. London: Institute of Latin America Studies.

Hauser, P.M. and Schnore, L.F. 1966. *The Study of Urbanization*. New York: John Wiley.

Instituto Nacional de Estadística. 2003. *Encuesta de Estatistica Nacional*. La Paz.

Lazar S. 2002. *Cholo Citizens: Negotiating Personhood and Building Communities in El Alto, Bolivia*. Doctoral Thesis. London, Anthropology Department: Goldsmiths College, University of London.

Napolitano, V. 2002. *Migration, Mujercitas, and Medicine Men: Living in Urban Mexico*. London: University of California.

Pérez de Castaños, M.I. 1990. *Derechos Humanos y Ciudadanos em El Alto*. La Paz: Defensor del Pueblo.

Perlman, J. 1976. *The Myth of Marginality: Urban Poverty and Politics in Rio de Janeiro*. Berkeley, CA: University of California Press.

Rengifo Vasquez, G. 1998. The Ayllu, in *The Spirit of Regeneration: Andean Culture Confronting Western Notions of Development*, edited by F. Apffel-Marglin and PRATEC. London: Zed Books, 89-123.

Revello Quiroga, M. 1996. *Mujer costumbre y Violencia en la Ciudad de El Alto*. La Paz: Cidem.

Sandoval, G. 1985. *La Ciudad Prometida: Pobladores y Organizaciones Sociales*. La Paz: ILDIS-SYSTEMA.

Urton, G. 1992. Communalism and Differentiation in an Andean Community, in *Andean Cosmologies Through Time: Persistence and Emergence*, edited by R. Dover, K.E. Seibold and J.H. McDowell. Bloomington, MD: Indiana University Press, 231-66.

Weismantel, M. 2001. *Cholas and Pistachos: Stories of Race and Sex in the Andes*. Chicago, IL: University of Chicago Press.

Chapter 12

A 21st Century Sustainable Community: Discourses of Local Wellbeing

Karen Scott

Today we need collective deliberation to do things that cannot be done in any other way. One is to forge some agreement about society's ultimate goals – what we mean by wellbeing, and how we are to achieve it. (Mulgan 1997: 191)

There is now an extensive literature covering a range of theoretical and empirical attempts to define and measure wellbeing, quality of life, life satisfaction, welfare, utility, happiness, human flourishing, human development and so on. This chapter is not concerned with fixed definitions but rather with dynamic discourses of wellbeing. It focuses on the concepts of wellbeing and quality of life as intuitively understood generic terms relating to a wide variety of ideas on what constitutes a 'well-lived life' (Dasgupta 2004: 13). Like all grand but slippery terms that have high rhetorical value in political discourse (freedom, justice and democracy to name a few others) they can be mobilized in different ways to support different agendas over time. Wellbeing is essentially a political concept and a focus on discourse explores how notions of wellbeing are constructed and used in different contexts, what and whose values they reflect, which accounts dominate, what is their impact and on whom.

The dominance of utilitarian theories of maximizing welfare[1] in western liberal democracies during the twentieth century has been documented by many. The satisfaction of individual preferences has been the major influence on wellbeing studies in economic theory for over a century and as a consequence political discourse has placed huge importance on raising GDP levels (for example see Dolan et al. 2006, Dasgupta 2004, Levett 1998, Veenhoven 1996). Over the last half century, large shifts in conceptualizing wellbeing have occurred and alongside this the concepts of place and space have been mobilized in various ways. Responding to increasing dissatisfaction with GDP as the predominant welfare indicator, the 'social indicators movement' in the US and Europe in the 1960s and 1970s promoted a multi-faceted view of quality of life underpinned by social justice ideals. This resulted in a proliferation of non-economic indicators covering domains such as health and education and leading to a profusion of work which mapped the spatial patterns of inequality at increasingly detailed

---

1   In utilitarian literature and in economics 'welfare' is synonymous with happiness, life satisfaction and 'utility'.

levels from the global to local. In addition, the environmental impacts of industrialization came to the fore in the 1980s producing a wide variety of discourses about human wellbeing in relation to the ecological integrity of the planet. The theory of sustainable development gained prominence as a way to balance economic, environmental and social pressures and the neighbourhood environment has often been the primary focus for action through subsequent Local Agenda 21 initiatives. In the last ten years or so, a new discursive wave has emerged driven largely by evidenced amassed in the economic and psychology disciplines about patterns and causes of subjective wellbeing. This has generated work which includes the psycho-social dimensions of individual wellbeing where relationship to place is a key theme.

In the United Kingdom, until very recently, the concepts of wellbeing and quality of life have been intertwined with government sustainable development discourses. For instance, sustainable development is often presented as 'quality of life for everyone, now and in the future' (DEFRA 2005) or 'social, economic and environmental wellbeing' (DETR 2000a). These discourses exist alongside public health discourses, which have tended to conflate wellbeing and health concepts (Atkinson and Joyce 2011, Ganesh and McAllum 2010, Fleuret and Atkinson 2007). In 2005, a greater commitment to explore and refine the concept of wellbeing was set out by national government (DEFRA 2005) and several studies were commissioned (for example: Dolan et al. 2006, Marks et al. 2006). The Whitehall Wellbeing Working Group developed a statement of common understanding regarding the definition of wellbeing as a positive mental state enhanced and supported by various social, environmental and psychological factors (DEFRA 2007). Despite striving for greater clarity regarding definition and measurement of wellbeing, struggles over the meaning of wellbeing represented 'stakes in the ground' for particular departmental positions or agendas (Ereaut and Whiting 2008: 13). Similarly, recent work shows how wellbeing is variously framed and constituted at the local governance level by different local authority partnerships (Atkinson and Joyce 2011).

Despite these various discursive framings of wellbeing and struggles over its definition, the policy gaze for improving wellbeing often focuses on the same place, the neighbourhood environment. For example, in the UK, the Local Wellbeing Project which developed the first set of local wellbeing indicators focuses heavily on 'place-shaping' at the neighbourhood scale as a key mechanism for improving wellbeing (Steuer and Marks 2008). The rest of this chapter will consider the case study of a contentious place-shaping project in north east England focussing on the discursive framings of wellbeing used by policy makers involved. It will end by drawing some conclusions about the impact of these particular discursive strategies on the project and the residents. Notions of participation and democratic debate are considered as part and parcel of wellbeing and place.

Creating a 21st Century Sustainable Community: National Context

In 2003 the 'Sustainable Communities: Building for the Future' Plan was launched by the New Labour government in the UK to direct a £38 billion programme of regeneration. This was hailed as a major initiative to promote economic development within a sustainable framework and had a strong focus on local community governance as a vehicle for guiding development (Raco 2005). This locality based 'sustainable community' was a central guiding principle for local government strategy, where the focus was on increased quality of life through safer, cleaner, greener, economically vibrant neighbourhoods. Through discourses of participation and 'active citizenship' neighbourhood communities were seen as key in solving local problems, where the focus was on the most disadvantaged (Jupp 2007, Taylor 2007a, Raco 2005, Fremeaux 2005). An explicit policy goal was the promotion of 'mixed communities' and changes to the planning system were proposed to facilitate the development of mixed tenure schemes which would include sufficient affordable housing to increase social inclusion (ODPM 2005). This reflects the widespread belief that a spatial concentration of poverty produces an additional negative 'neighbourhood effect' on a person's life chances and that increasing social mix will have a number of advantages including greater sustainability and social cohesion.

However, recent reviews and critiques of this approach argue that this belief is not well substantiated by the evidence and there is a lack of research on post development impacts on residents (Tunstall and Lupton 2010, Cheshire 2007). Furthermore, scholars point to the false dichotomy between *compositional* and *contextual* approaches where spatial inequalities are seen on the one hand as purely the result of aggregated individual characteristics or, on the other, the result of some additional 'area effect' (Taylor 2007b, Castro and Lindbladh 2004). These debates connect in a complex way to substantive evidence about place, wellbeing and deprivation. For example, research on the development of place attachment in mixed communities shows a complex interaction of factors including social relationships, sense of security, length of stay, age profile and personal 'fit' (Livingston et al. 2008). There is a need to understand in much more depth the relationship between wellbeing, place and deprivation in the context of a mixed community policy approach.

This chapter focuses on how these relationships were understood by policy actors who were responsible for a controversial regeneration project, which involved the proposed demolition of a social housing area to create a sustainable mixed community. The author conducted ethnographic research for three years (2004-2007) in a local authority and used discourse analysis to investigate the discursive context in which policy and practice takes place. The research included interviews with policy actors, focus groups with residents and attendance at consultation events, steering group and resident meetings. The analysis focused on an investigation of how agents and entities are constructed and the key assumptions and metaphors present in the shared language of the policy actors

(Dryzek 1997). This concept of discourse is focused on the actor as an agent using language intentionally in 'discursive strategies' (Rydin 2003: 49) where language is 'determinate' and can influence thoughts and actions and therefore policy (Castro and Lindbladh 2004: 261). The quotations presented here are drawn from interviews with policy actors who were directly involved in the regeneration strategy and decision making, as well as local residents. Due to the sensitive nature of some of the material, specific roles or affiliations of respondents have not been identified.

Bates Colliery Regeneration, Blyth, Northumberland

Blyth is a small town on the north east coast of England with a population of 35,500. It was once a centre of excellence in shipbuilding and in the 1960s was the busiest coal port in Europe. The terminal decline of both the shipbuilding and coal mining industries meant that by the 1990s Blyth was suffering economic disadvantage, social deprivation and environmental degradation. In 2000 a regeneration body, South East Northumberland and North Tyneside Regeneration Initiative (SENNTRi),[2] was formed to create a 'socio-economic step change' in the wider area with a key aim to 'improve quality of life'. The board of SENNTRi included local authority, Regional Development Agency (RDA) and English Partnerships[3] representatives. In 2002 they published their 'Corridor of Opportunity' prospectus for linking this post-industrial sub-region, which included Blyth, with the city region of Tyne and Wear to generate economic opportunities and investment. Indicators of success included jobs created, amount of private sector investment, number of residential units and amount of commercial floor space created. These fed into regional indicators of economic activity such as Gross Added Value (GVA)[4] statistics.

Working closely with Blyth Valley Borough Council (BVBC),[5] SENNTRi commissioned a study to look at regeneration potential along Blyth estuary, an area including large amounts of ex-industrial land. The Blyth Estuary Development Framework Plan published in January 2005 outlined 'a range of opportunities available to unlock the potential of the sub-region and transform the area into a successful, progressive and sustainable community with a strong local identity' (Llewellyn Davies 2005). A derelict and contaminated deep coal mine site next

---

2   SENNTRi ceased to exist in 2009 and was replaced by the South East Northumberland Regeneration team, which is fully funded by Northumberland County Council and carries out the regeneration roles previously delivered by former district authorities and SENNTRi.

3   The UK government's national regeneration agency.

4   GVA is a measure of economic activity which contributes to the estimation of GDP.

5   Due to a Local Government Review by central government in 2007, the district council Blyth Valley Borough Council (BVBC) ceased to exist from April 2009 and was incorporated into the much larger unitary authority, Northumberland Council.

to the river, Bates Colliery, was identified as a potential residential and retail development 'flagship' for this transformation. The site was described as 'an opportunity to create a regenerated sustainable community for the 21st century as a pivotal component of the wider renaissance of South East Northumberland and North Tyneside region' (SENNTRi, 2006).

The proposals included the demolition of an area of social housing, Hodgsons Road Estate, located adjacent to the site. The estate consists of 292 households, with approximately 15 per cent of the estate having been sold to tenants through the 'Right to Buy' scheme. The majority of homes are interwar two and three bedroom houses with some bungalows and flat-roof properties being added in the 1960s and 1970s. Some of the houses are very poor quality and about 80 properties on the estate do not meet the Decent Homes Standard.[6] During the 1990s the estate developed a poor reputation for anti-social behaviour and properties were difficult to let.

Hodgsons Road Estate is in Croft ward, which is amongst the most deprived 10 per cent of wards in the country in terms of income, education, health and child poverty levels (Noble et al. 2008). However the estate is separated from the rest of the ward by a main road and seen as a rather isolated enclave of more severe deprivation. A survey carried out on the estate by the council in 2005 showed that there was a higher than average older population on the estate and that 40 per cent of all households contained a person or persons with a serious medical condition. The new development was presented as an opportunity to address the poor housing on the estate, integrate the area better with the town and increase quality of life for residents. In November 2005, residents were informed of the proposals by letter, invited to attend a consultation event and consulted individually by a team of officers. This marked the beginning of (up to the time of writing) five years of uncertainty for the residents. The rest of the chapter gives an account of some aspects of the development proposals, the consultation and the impact upon some of the residents. It discusses how wellbeing and quality of life was conceptualized in place-making discourses by key policy actors and describes the influence of new discourses of subjective wellbeing.

Transforming Tomorrow

It is important to note that the demolition of Hodgsons Road Estate was seen as key to the Bates Colliery development which in turn was viewed as a catalyst to the wider socio-economic transformation of Blyth. The development was described consistently in promotional documents as a 'flagship', 'beacon', 'visionary' or 'landmark' development which would allow the 'opportunity' for 'releasing untapped potential' to 'flow' out and effect a 'step-change' towards a 'sustainable' future and 'a better quality of life'. This discourse of transformation

---

6   A UK national guideline of what defines a decent standard of housing following the Green Paper 'Quality and Choice: A Decent Home for All' (DETRb 2000).

through large scale physical regeneration was used and promoted by SENNTRi and senior council officers and had an important rhetorical purpose. The site had complex land ownership issues and necessitated negotiations with seven different private and public sector landowners. A strong, positive and coherent vision was key in motivating all stakeholders into a joint master-planning process, to realize the vision and avoid piecemeal development of the site. Similarly, a proposal which involved demolition of properties is always controversial and likewise a strong positive message of the benefits to residents was needed. The discourse of transformation however, was not purely rhetorical but reflected deeply held beliefs by key policy actors that substantial physical regeneration would lead to economic growth and that this was central to improving wellbeing for disadvantaged people in the borough:

> So new houses, new facilities, new infrastructure, new schools ... the success for me is to look at some of our problems and, rather than just dipping our toe in the water, taking everyone along and going for the radical change, the transformational change ... not just for a few people who live around there at the time but for future generations as well. (Policy officer)

> If you were to change the housing and suddenly they've got something which is brand new ... that would lift their self-confidence and their whole demeanour and that would hopefully transfer through to them being able to then say, right well OK I'll go and have a look for a job. (Policy officer)

The creation of 'mixed communities' was an explicit policy goal to attract more affluent people into Blyth by providing 'executive type housing' and simultaneously dilute concentrated areas of deprivation by interspersing social housing amongst these more affluent areas, or 'pepper-potting poverty' as one officer called it. Importantly, a key idea mobilized in these discourses is that development would not only change the economic profile but also bring '*a different type of person*' into the borough:

> We have the opportunity in Croft and across the waterfront in Blyth to bring in people with a different perspective. I think that's what will change the culture of Croft ward, not small scale projects and interventions, I think they're helpful but they ain't going to change Croft ward and if you don't change the nature of the people then Croft ward will still be in the bottom 10% of wards in seven years' time. (Policy officer)

The idea of attracting affluent, aspirational and entrepreneurial people to improve the culture of Blyth went hand in hand with improving the image of Blyth. Bates Colliery would become 'an attractive gateway to Blyth' with a 'unique sense of place' (John Thompson and Partners, 2007) and in these discourses Hodgsons Road Estate was often constructed as a visual blight:

> Hodgsons Road is a an eyesore site, so if we're wanting to create change in Blyth and allow it to punch its own weight within the region those kind of eyesore sites need to go. I might be sounding a bit hard here but you know there are people that live there and they need to be dealt with but that's not our problem, that problem is Blyth Valley's, that's a local authority housing development. (Policy officer)

The high statistical levels of deprivation on Hodgsons Road Estate meant that it tended to be viewed as an homogenous entity which was causing a cultural and visual barrier to quality of life, not only for the residents who lived there but also for the rest of Blyth and for potential residents who might want to live in a high quality development next to the river. In these discourses, wellbeing was a vague socio-economic concept, often conceptualized as the inverse of deprivation. It was contrasted with a particular construction of deprivation in which 'culture' was often mentioned as both cause and solution. The physical state of the estate was seen as a manifestation of this culture and the spatial strategy of mixed communities put forward as a solution.

Some implicit links between wellbeing and place were consistently presented throughout these discourses: first, that a new house will have a transformational effect on the disadvantaged individual, will lift their confidence and lead to better employment chances; secondly, that an attractive environment will draw more affluent people into the area, improve the local economy and that will have knock on benefits to the disadvantaged, through increased job opportunities; thirdly, that the spatial strategy of 'pepperpotting poverty' will raise the aspirations of the disadvantaged by self-comparison to wealthier neighbours, leading to more entrepreneurial activity. The key thread is that attractive physical regeneration will lead, directly or indirectly, to better wellbeing through economic success. As such place-shaping strategies were focussed predominantly on increasing the quality of the material and visual realm in order to attract a different type of person or encourage different behaviours.

However, when questioned about the detailed pathways by which wellbeing would be improved *and for whom*, many policy makers were unconvinced by the same discourses that they were promoting. Attracting wealthier people would reduce statistical levels of deprivation in a particular area and improve economic growth indicators but it was unclear in what ways this would reduce individual deprivation:

> I will accept is that the concept of trickle-down does not really work because what's happened nationally is that economic success has tended to result in either two incomes or no incomes. (Policy officer)

> This was originally identified as a private sector regeneration … the private sector do the planning application, the private sector build whatever they want

to build there, we'll get a few quid out of them for benefits to the community I
don't see how that provides a benefit for the current community. (Policy officer)

In addition, some policy makers, whilst wanting to get rid of 'poverty ghettos'
were also concerned that if deprivation was scattered throughout the borough
in mixed communities then it becomes more invisible. This makes ward based
deprivation statistics look better but makes it harder to argue for nationally
distributed intervention funding for the most deprived areas.

Hodgsons Road Estate

In November 2005, an initial consultation event for residents was held at a local
venue, The Comrades Social Club. There were large display stands from SENNTRi,
the council and a housing association, all projecting images and messages of a
'better quality of life' and 'transformation' in Blyth. A team of council officers
were there to consult with residents but two officers present expressed cynicism
about the consultation and described it as a 'marketing exercise' designed to 'bring
the community round'. Some residents were annoyed at what they perceived to
be glib promises and a lack of recognition of the upheaval that the development
would cause:

> The first meeting we had at the Comrades I turned round to one of the people on
> the council team and I says, 'So do you think it's worth going to see a solicitor?'
> and I'm not joking his face nearly hit the floor, he gans, 'What d' you wanna do
> that for?' … It was just like well, that's for your house, that's for your house,
> that's for your house, you can move in that one, you can pick your own house,
> you can live next to your neighbours if you want and then we'll pull them all
> down, and we'll build all the houses up and everybody's happy, and it's like no,
> it's not, it's not as simple as that. (Resident's focus group)

In addition, as discussed above, Hodgsons Road Estate was generally constructed
as a homogenous entity of deprived people and therefore, problematically, some
policy makers felt they were consulting with the very culture that they wanted to
change. This led to an instrumental view of consultation by some:

> I think the consultation has been necessary but the redevelopment of the area, in
> my mind, the priority is about bringing new people into Blyth. (Policy officer)

After this initial consultation, a team of council officers conducted an in depth
survey with each household. A database of survey responses was set up by
the council officers. This included personal information about the residents'
circumstances and their opinions about their houses and the neighbourhood. A
significant message that came out of this consultation was that many people had

strong attachments to their neighbours and were concerned about how the proposals would affect those relationships. They also felt that the estate had been forgotten by the council and in recent years had gone downhill. Most of the dissatisfaction seemed to concentrate on the local area due to the physical shabbiness, lack of amenities and anti-social behaviour. Although many residents were dissatisfied with their houses, this centred on the windows, kitchens and bathrooms which were in poor condition and in need of refurbishment. Partial demolition or renovation was not presented as an alternative to the residents (although this had happened elsewhere in the borough) because council officers felt that this option would be more expensive in the long term as it would not address wider issues or create the cultural and economic step change needed.

A majority of the householders on the estate were broadly in favour of the proposals but the range of responses varied in nature. Some people were keen to move away from the estate but had not previously been able to sell their house or get a transfer and saw the development as an opportunity to leave. Others would have been happy with new kitchens and bathrooms and agreed to demolition only if there were guarantees to stay with their neighbours in the same area. Some agreed but were concerned about increased rent levels. Some of the older residents were worried about how long the plans would take and the impacts on their health:

> I don't want to start moving when I'm seventy three … if they were to buy it off us, even if it's a couple of thousand below the going rate, I would take it, to get out and soon get settled again rather than wait five years and then get stretchered out of the place with the removal van. (Resident's focus group)

Some home owners on the estate were against the plans from the start. They felt that the compulsory purchase arrangements of market value plus 10 per cent offered by the council for their houses would not be adequate to cover the financial and emotional investment they had put into their house. Some had been able to buy their property cheaply under the right-to-buy scheme and had since paid off their mortgage and they were worried about having to take out another mortgage.

In short, the consultation showed that there were a variety of people in Hodgson's Road Estate with a variety of preferences in regard to the proposals. Although there are high levels of disadvantage, the estate is not homogenous and objective statistics do not reveal the mix of people and the wide range of living standards. The estate consists of several small communities who can mistrust one another but within these small groups, support can be very strong, with neighbours performing substantial amounts of unpaid care for each other. Despite the diversity of views on the estate and that the consultation seemed rather prescribed, a majority agreement for the demolition by residents gave policy actors legitimacy to go ahead with the plans.

In July 2006, the council employed an urban design consultancy from London to develop a master plan of Bates Colliery. The steering group consisted of representatives from the council, SENNTRi, English Partnerships, the RDA and the three largest private sector landowners. In addition a working group was set up which consisted of other interests including environmental agencies, community development workers and resident representation (although the designated resident rarely attended). The consultants met with both groups separately every six weeks or so but it was clear that the steering group held the decision making power.

In September 2006 a highly publicized community consultation weekend was held by the consultants to elicit residents' views on design and planning options for the site. A very low number of households (less than 20) were represented at this consultation. Nevertheless residents present expressed a range of views and concerns, often conflicting with each other. For instance, there was a lively discussion regarding different housing styles. The consultants took note of all contributions and fed these back meticulously and in detail at the next steering and working group meetings. It is unclear how all these diverse and conflicting views (by a minority of residents) were then fed into the master-planning process. Some resident preferences for a more traditional house design (similar to what they already had) were explicitly discounted by policy actors in steering group meetings on the grounds of residents having 'a lack of education'. Ideas of 'a sense of place' or 'distinctiveness' were predominantly confined to the visual realm and seemed largely to reflect the design preferences of the professionals involved.

There was a hiatus in the process as detailed market testing was carried out to explore the economic feasibility of various options for the site. In December 2006 regular consultation with the residents petered out. Although a Strategic Development Guide was produced in July 2007 to inform planning applications for the site, between 2008-2010 the development proposals were enmeshed in several complex issues including changes in the political and administrative landscape following the local government review in 2007 and local elections in 2008. Over the course of time both tenants and owner occupiers became increasingly frustrated:

> This has been going on for two and a half year now.

> Well I tell you what it is, it's we're thinking shall we do the painting, shall we do the decorating? Do we bother? We just don't know what's happening.

> This guy can't sell his house for love or money because the council have said they are going to knock it down.

> It's like we're in limbo. (Comments from four residents – focus group)

> The initial goodwill's been lost.

> This is affecting lives more and more because of the uncertainty.

> What we want to know is, is it or is it not going to happen.

It doesn't take much on this estate for people to get dejected, it's like a balloon the longer it goes on the more it deflates. (Comments from four residents – resident's meeting)

One man described in detail the impact it had had on his life:

It got to the point, me and my wife a while back got very close to splitting up … a lot of it was depression caused by this pushing it over the top. This regeneration thing has caused really big problems particularly with her, basically, she ended up on tablets for depression because of it. A lot of people might say 'oh that's a bit drastic' but everybody's different. Like I say we're not particularly big, ambitious people, we started off getting married, she was 16, I was 18, we had nothing, we lived in a little flat, luckily we got a house in Hodgsons Road, lovely house, lovely area, lovely neighbours, eventually we bought it and then I came into some money and we decided to do it up, not to sell it, but to improve the quality of life, nice little conservatory, sit in the sun, redo the kitchen, got the plans done, paid the deposit for the conservatory, then this letter lands on the doorstep … It has caused a lot of problems, a lot of problems.

Discussion

A dominant discourse of transformation in Blyth was driven by ideas of economic growth through physical regeneration. It mobilized a particular concept of a sustainable community, that of a 'mixed community' where a visually attractive place and aspirational people were the key components of success. Wellbeing and quality of life were rather under-developed concepts which could be achieved as the result of reducing socio-economic deprivation through place creation. Objective indicators of deprivation were used to argue for the wholesale demolition of Hodgsons Road Estate, to remove an eyesore and to change a culture. Yet it was unclear to most officers in SENNTRi and BVBC how this large scale physical regeneration would benefit those most vulnerable or reduce social inequality and they struggled to evidence or articulate this. The discourse of economic growth through place creation held little real explanatory power for policy makers in terms of addressing individual deprivation as opposed to making objective ward indicators look better.

Whilst there were obvious and important potential benefits involved in the regeneration plans, and deprivation that demanded action, there were also actual and potential negative impacts and these were under-represented and under-discussed. Justification for the demolition was articulated in quality of life discourses and evidenced by the high levels of deprivation and infirmity on the estate, yet ironically very little attention was paid to the impact of the planning and implementation process on the wellbeing of these same, presumably vulnerable, residents. This was partly due to the promotion of a very positive and dominant transformation discourse where the focus was on a vague future vision of quality of life. Ill-defined notions of wellbeing can serve as a diversion making it difficult to explicitly discuss theories of cause and effect and what the solutions might be (Ganesh and McAllum 2010, Seedhouse 2001).

Although many resources were committed to resident consultation at the start, support was gradually withdrawn over time as resources were needed elsewhere and despondency set in producing a further negative effect on wellbeing as residents became cynical and distrustful. The initial community consultation gave policy actors legitimacy for the demolition decision because a majority of residents had been in favour. This became a tablet of stone which was used as evidence time and again that the council were acting on the community wishes. However, the community was not homogenous. Nor did their views remain static. Large, heavily publicized, one-off consultation exercises were focussed on collecting and collating views rather than debating conflict. The pathway from consultation to policy decision was obscure and as such a variety of views could be filtered out with little transparency. The steering group for the development was predominantly concerned with the negotiation between private sector interests and the macro socio-economic needs of the area, the latter evidenced by deprivation statistics. This group was inaccessible to residents and removed from the complexity of their lives and the conflicts between their views. As many scholars (for example Cahill 2007, Cooke and Kothari 2001, Laessoe 2007) have pointed out, the tendency for participation to focus on consensus rather than difference can lead to over simplistic accounts, as was the case here.

On a more positive note, new discourses of subjective wellbeing which started to become influential towards the end of the research period allowed some policy makers to engage more deeply with a more complex notion of wellbeing. As the project progressed, national interest in subjective wellbeing was increasing and new evidence around wellbeing was becoming more publicly discussed. In addition, having a researcher present in the council opened up debate about definitions and measurement of wellbeing and the resulting contestation was influential. For some key officers a far greater reflection about wellbeing and quality of life ensued. Policy makers started quoting from a different evidence base concerning subjective wellbeing. For example, one key aspect of wellbeing which the new discourses highlighted is its relational nature, namely that it is relative rather than absolute income which matters and that social comparisons have the most effect on satisfaction and happiness (Michalos 1999). Therefore the strategy of creating 'mixed communities' may actually decrease subjective wellbeing for disadvantaged people who may compare themselves more negatively to their affluent neighbours. There was also a greater acknowledgement of the negative impacts of the regeneration process and the need to pay more attention to psychological wellbeing:

> Once we'd learnt about these things [the high rate of older/disabled people on the estate] we didn't organize an intervention action plan to address a number of

issues on the estate particularly support in terms of disability/old age. We did go to other places, to Stockton, and talk to them about the process but it was mainly about the compensation package not about managing uncertainty. I think this is a gap in physical regeneration. It's something we can definitely learn from. (Policy officer)

There are things we don't capture because we don't ask these people enough in a structured way or a regular way how they feel, how they're getting on, are they engaged, are they confident, how well do they feel. (Policy officer)

In November 2007, several BVBC officers attended a high profile conference on wellbeing which included high profile keynote speakers Nic Marks from the New Economics Foundation and Martin Seligman, from Pennsylvania University, both influential figures in the new discourses of subjective wellbeing. The conference caused much debate in the council and a more nuanced view of wellbeing began to be formalized in strategic policy documentation produced in 2008:

In some areas, particularly economic development, wellbeing measures tend to conflict with more traditional regeneration measures and in some cases, with sustainability measures ... The search for increased productivity can conflict with measures of wellbeing. (Blyth Valley Borough Council 2008)

This generated greater reflexivity amongst some regarding the potential negative effects of current place-making strategies underpinned by narrow conceptions of wellbeing, where the idea of 'place' was conceived in rather limited terms. A focus on general wellbeing, rather than improving objective indicators of deprivation, opened up new discussions. Importantly, a more nuanced view of wellbeing developed which allowed policy makers to question who would and would not benefit from current policy strategies like the mixed communities approach, instead of promoting an 'everyone wins' logic. A greater sensitivity to how terms like 'quality of life' and 'wellbeing' were used to justify spatial strategies ensued. For some, their idea of wellbeing shifted from a narrowly defined socio-economic concept to one which included psycho-social elements like place attachment and sense of belonging. Importantly, this started to pave the way for complex discussions about what wellbeing is, whose wellbeing should take priority and what the local authority had responsibility for. In turn, this led to a deeper recognition of the inherently political nature of the concept and therefore the importance of defining wellbeing locally through deeper consultation and engagement of partnerships and communities, where democratic contestation is pivotal. The claim that making wellbeing the focus of policy can remind us what policy is trying to achieve (Fleuret and Atkinson 2007: 115) resonates with this study. A strong focus on participation and debate should underpin a 'communication-centred notion of wellbeing' (Ganesh and McAllum 2010: 496) which can guide place-shaping strategy.

Note

Researcher involvement ended in 2007. In August 2010 the new unitary authority Northumberland Council initiated a further master-planning process due to the changed political and fiscal landscape and, at the time of writing, three partial demolition options have been produced for Hodgsons Road Estate which include the retention of a large proportion of owner occupied housing along the main routes and 70 per cent affordable housing in the new development. The residents have informed the council that they do not wish to be consulted further until there is a firm proposal for development, therefore a communications protocol has been adopted to try to manage all communication with residents.

Acknowledgements

Many thanks to Sarah Atkinson, Derek Bell, Menelaos Gkartzios and Nicola Thompson who provided useful comments on the draft.

References

Atkinson, S. and Joyce, K. 2011. The place and practices of wellbeing in local governance. *Environment and Planning C*, 29(1), 133-48.

Blyth Valley Borough Council. 2008. *Blyth Valley Strategic Objectives 2008-2025*.

Cahill, C. 2007. Well positioned? Locating participation in theory and practice. *Environment and Planning A*, 39, 2861-65.

Castro, P.B. and Lindbladh, E. 2004. Place, discourse and vulnerability – a qualitative study of young adults living in a Swedish urban poverty zone. *Health and Place*, 10, 259-72.

Cheshire, P. 2007. *Segregated Neighbourhoods and Mixed Communities: A Critical Analysis*. York: Joseph Rowntree Foundation.

Cooke, B. and Kothari, U. 2001. *Participation: The New Tyranny?* London: Zed Books.

Dasgupta, P. 2004. *Human Well-Being and the Natural Environment*. Oxford: Oxford University Press.

DEFRA. 2005. *Securing the Future: Delivering UK Sustainable Development Strategy*. Norwich: The Stationary Office.

DEFRA. 2007. *A Common Understanding of Wellbeing* [Online]. Available at: http://www.defra.gov.uk/sustainable/government/what/priority/wellbeing/common-understanding.htm [accessed: 12 December 2010].

DETR. 2000a. *Local Quality of Life Counts: A Handbook for Menu of Local Indicators of Sustainable Development*. London: HMSO.

DETR. 2000b. *Quality and Choice: A Decent Home for All- The Housing Green Paper*. London: Department of the Environment, Transport and the Regions.

Dolan, P., Peasgood, T. and White, M. 2006. *Review of Research on Personal well-being and Application to Policy Making.* London: DEFRA.

Dryzek, J.S. 1997. *The Politics of the Earth: Environmental Discourses.* New York: Oxford University Press.

Ereaut, G. and Whiting, R. 2008. *What do We Mean by Wellbeing and Why might it Matter?* Linguistic Landscapes Research Report no. DCSF-RW073 for the Department of Children, Schools and Families.

Fleuret, S. and Atkinson, S. 2007. Wellbeing, Health and Geography: A critical review and research agenda. *New Zealand Geographer*, 63, 106-18.

Fremeaux, I. 2005. New Labour's Appropriation of the Concept of Community: A Critique. *Community Development Journal*, 40(3), 254-74.

Ganesh, S. and McAllum, K. 2010. Well-Being as Discourse: Potentials and Problems for Studies of Organising and Health Inequalities. *Management Communication Quarterly*, 24(3), 491-98.

John Thompson & Partners. 2007. *Bates Colliery Strategic Development Guide.* Report prepared for South East Northumberland North Tyneside Regeneration Initiative, English Partnerships and Blyth valley Borough Council.

Jupp, E. 2007. Participation, local knowledge and empowerment: Researching public space with young people. *Environment and Planning A*, 39, 2832-44.

Laessoe, J. 2007. Participation and Sustainable Development: The Post-ecologist Transformation of Citizen Involvement in Denmark. *Environmental Politics*, 16(2), 231-50.

Levett, R. 1998. Sustainability Indicators – Integrating Quality of Life and Environmental Protection. *Journal of the Royal Statistical Society Series A*, 161(3), 291-302.

Livingston, M., Bailey, N. and Kearns, A. 2008. *The Influence of Neighbourhood Deprivation on People's Attachment to Places.* York: Joseph Rowntree Foundation.

Llewellyn Davies. 2005. *Blyth Estuary Framework Plan.* Report prepared for SENNTRi.

Marks, N., Thompson, S., Eckersley, R., Jackson, T. and Kasser, T. 2006. *Sustainable Development and Well-being; Relationships, Challenges and Policy Implications.* Defra Project 3b. London: New Economics Foundation.

Michalos, A. 1999. Reflections on Twenty-Five Years of Quality of Life Research. *Feminist Economics*, 5(2), 119-23.

Mulgan, G. 1997. *Connexity: How to Live in a Connected World.* London: Chatto & Windus.

Noble, M., McLennan, D., Wilkinson, K., Whitworth, A., Barnes, H. and Dibben, C. 2008. *The English Indices of Deprivation 2007.* London: Department for Communities and Local Government.

ODPM. 2005. *Planning for Mixed Communities: Consultation Paper* [Online]. Available at: www.communities.gov.uk/archived/publications/planningandbuilding/planningmixed [accessed: 6 December 2010].

Raco, M. 2005. Sustainable Development, Rolled-out Neoliberalism and Sustainable Communities. *Antipode*, 37, 324-46.

Rydin, Y. 2003. *Conflict, Consensus, and Rationality in Environmental Planning: An Institutional Discourse Approach.* Oxford: Oxford University Press.

Seedhouse, D. 2001. *Health: The Foundations for Achievement.* Chichester: John Wiley.

South East Northumberland North Tyneside Regeneration Initiative, English Partnerships (SENNTRi). 2006. *Bates Colliery Proposed Residential Site Strategic Development Guide.* Blyth.

Steuer, N. and Marks, N. 2008. *Local Wellbeing: Can We Measure It?* London: New Economics Foundation.

Taylor, M. 2007a. Community Participation in the Real World: Opportunities and Pitfalls in New Governance Spaces. *Urban Studies*, 44(2), 297-317.

Taylor, M. 2007b. *Transforming Disadvantaged Places: Effective Strategies for Places and People.* York: Joseph Rowntree Foundation.

Tunstall, R and Lupton, R. 2010. *Mixed Communities: Evidence Review.* London: Department for Communities and Local Government.

Veenhoven, R. 1996. Happy Life-expectancy: A comprehensive measure of quality-of-life in nations. *Social Indicators Research*, 39, 1-58.

Chapter 13

'We are the River': Place, Wellbeing and Aboriginal Identity

Lorraine Gibson

> I was reared up on the river, that's where we get our name from. Barka means 'river', Barkindji means 'river People'. Without the river we lose our culture, we lose our identity. (Badger Bates, Barkindji Aboriginal man, personal communication 2003)

Improving health, living conditions and quality of life for families, individuals and communities are key criteria used by the Australian government in conceptualizing and measuring wellbeing (Australian Bureau of Statistics 2001). These three areas are stated to be a 'major driving force in human activity' and are closely linked to the broader social measures of economic performance and the state of the environment. However, different cultural notions of these criteria sees wellbeing experienced, interpreted and measured differently. When notions of wellbeing and place are linked, the stage is set for multiple misunderstandings.

Place is a social concept and the ways in which a physical place or space is understood, experienced and imagined is tied to cultural values and beliefs: 'only when space becomes *a system of meaningful places* does it become alive to us' (Norberg-Schulz 1969). It is 'culture', in the anthropological sense of a people's way of life, that constructs and guides these meanings, the attached values and the resulting emotions and behaviours in relation to place (as this is culturally understood) comprising, as it does, both physical and imagined factors. If an individual or group's sense of place or sense of the physical space to which an individual or group has attached particular cultural beliefs, feelings and ways of knowing is contested, denied, or indeed despoiled by others, what might this mean for wellbeing? This chapter provides an ethnographic account of the cultural significance of the Darling River (hereafter, 'the River') which flows through Wilcannia in far western New South Wales (NSW), Australia. It explores the ways in which Barkindji Aboriginal ways of knowing and experiencing the River intersect with, and vary from, those of Australian settler society and what this means for Aboriginal cultural health, cultural identity and wellbeing.

In general terms we can say that water is important for all people. Water is important whether or not an individual or group is necessarily cognizant of having any particular spiritual, cultural, economic or political interest in it. Simply stated, nothing lives without water. Depending on cultural predispositions and values there are of course differences in how water, whether a body or in a glass, is imagined and thought about. A vial of 'holy water' from Lourdes to which people attribute special healing properties, the water of the Ganges and the water flowing over a weir in Australia are at once the same but different. A practicing Catholic who is wheelchair-bound and seeking a cure may invest the Lourdes water with quite different meanings from the Hindu or non-believing Catholic who do not share their beliefs. Similarly, a practicing Hindu will invest the meaning of immersion in the Ganges quite differently to a practicing Catholic. There are, then, inter- and intra-cultural differences that touch on the different spiritual, economic and political meanings of water. These, in turn, will affect the ways in which we come to know place and the experiences through which we come to understand place (Relph 1976) and, how these in turn translate into perceptions and experiences of wellbeing as states of physical, mental or cultural health.

The small country town of Wilcannia has a shifting population of around 600 people of whom the large majority (conservatively, 68 per cent) are Aboriginal.[1] For the Barkindji people who are the traditional Aboriginal owners of the land around Wilcannia, the Darling River is known as the *Barka*. Indeed, 'Barkindji' literally means 'belonging to the River' (Hercus 1993). However, the different cultural meanings of the River and water float and are fluid depending not least upon power relations.

Since the British colonization of Australia in 1788 water – its plenitude and lack – has been embedded in ambivalent and complex ways within the Anglo-Australian imaginary. Across much of Australia, land use change by Europeans has seen a cumulative and increasing impact upon the arid and semi-arid areas of the far west of NSW. Since the 1860s, cattle and sheep grazing and the resulting vegetation loss and soil degradation has altered the ecology of the Darling River and its environs. Variations to water flows through policy regulation, damming, the demands of 'up river' irrigators, the introduction of non-native fish species, and, more recently, long periods of drought has altered the quality and quantity of water, as well as altering visual aspects of the surrounding country. Erosion, increasing salinity, pollution, weed invasion and changes in native flora and fauna are just some of the effects. Aboriginal and non-Aboriginal experiences of these changes and understanding of their causes and effects are, however, often discrepant. The differences between Aboriginal and non-Aboriginal ways of *thinking about* and *experiencing* the River elucidate the gap in these relationships. Indeed difference, relative difference, the spaces between and the ways in which these connect to one's sense of wellbeing and place comprise the broader scope of this chapter.

Within a day of my arriving in Wilcannia where I lived for 16 months undertaking anthropological fieldwork, I was introduced by an Aboriginal man as having come to Wilcannia to 'study the River'. Almost immediately I was told by many Aboriginal people, 'We are Barka', meaning in the literal sense, 'We are the

---

1   Exact figures are difficult to obtain as many Aboriginal people are highly mobile and spend varying amounts of time living with and visiting their wider kin networks across NSW and beyond. Many are also either reluctant or uninterested in completing the Census.

River'. People expressed that 'the River is everything' and 'without the River, we are nothing, we have no identity'. On the face of it, such explicit responses are to be expected since the River was the stated nature of my research enquiry.

There is, of course, much about cultural identity that is not, and indeed cannot, be verbalized, and which 'goes without saying' (Bloch 1992). Bloch uses this phrase to outline the disparity between language and thought; to demonstrate that unlike language, everyday thought relies on 'clumped networks of signification' which are 'only partly linguistic', including as they do the integration of 'visual imagery, other sensory cognition, the cognitive aspects of learned practices, evaluations, memories of sensations, and memories of typical examples' (Bloch 1992: 120, 130). My point here is not to trivialize verbal expressions of cultural identity as being of little consequence. Bloch's differentiation between thought, linguistic expression and experience is precisely what makes Barkindji's explicit and repeated forms of identity important. If a people identify as 'being the River' and 'belonging to the River' and they also perceive the River as being 'Barka in Trouble' (as one artist titled his painting), then in what ways is this identification being imagined, worked through, thought about? What does this mean for the Barkindji people, for culture, wellbeing and identity?

These, and other unsolicited verbalized forms of identity, were constantly repeated throughout my time in Wilcannia – amongst Barkindji, to other non-Aboriginal residents and non-residents, to visiting government and non-government agencies and to providers who serviced or had dealings with people in town. The expressions 'We Barka' and 'We Barkindji' were often used interchangeably and were said for all sorts of reasons, at different times, and in different contexts. They were expressed, for example, when relaying the connection between being Barkindji and the practice of eating fresh fish and johnnycakes cooked in the ashes of a fire; when talking of liking – indeed, needing – to go fishing in order to maintain some peace of mind; when expressing that to be Barkindji means needing contact with the River through, for example, camping along the riverbank with family, with friends or in isolation. Strong connections are also drawn between what it is to 'be Barka' or to 'be Barkindji' and the frequency and incidence of townspeople who go in and out of jail and the propensity for fighting and jealousy in relationships between men and women. In a drinking context, I was often told, 'You with Barka now', 'You have to be like Barka', 'Enjoy yourself', and 'Be like us', which translates to the necessity to 'get charged up', to drink until drunk. These expressions are often clearly and directly place- or River-related, while at other times they do not directly connote a River association. What it is to is to be Barkindji or Barka – as expressed and, indeed, as practiced – is, importantly, a holistic identifier, which encompasses all of life as lived; it cannot be reduced to the physical or spatial place of the River in any isolated or separated sense. Place here is cognized and experienced; it is reflexively seen, interpreted, accorded values and is subject to cultural conditioning: there is both conscious and unconscious awareness of place and one's place in it (Tuan 1980, 2005): culture and place are intrinsically linked.

The communal Barkindji catch-cry of 'Being the River' has, however, little resonance for most whites in Wilcannia. Although a few local whites speak rhetorically about the importance of the River for the Barkindji people, most consider the group expression of being 'People of the River' as having little ontological meaning in Aboriginal cultural terms. A great deal of white ambivalence towards Barkindji claims of a special cultural relationship with the River rests on wider public perceptions of what constitutes 'Aboriginal culture' and who is thought to possess it. Because both black and white senses of place are tied to and endowed with cultural values, and because the hegemony of white discourses and meanings often hold sway in inscribing and controlling what is seen to constitute 'Aboriginal culture', Barkindji cultural values are often overridden in ways which are damaging to Barkindji agency and, therefore, wellbeing.

For most of Wilcannia's whites, Aboriginal people in town are seen to be similar to themselves in many ways. Both share the same geographic space, live in houses, shop at the one local supermarket, go to the local club, watch television, videos and DVDs; they talk about the River, about water (or the lack of it), about fishing and about fish (or the lack thereof). These are areas which, on the surface, seem similar and overlapping. Ways of life seem to converge, yet there are differences in interpretation and understanding of these daily experiences and goings-on. In fact, the *appearance* of similarity only serves to fuel misunderstanding and anger when Barkindji assert cultural rights and beliefs in relation to the River and, importantly, its resources in ways which deviate and are at odds with those of the dominant culture. Most whites are unwilling and or unable to grant Barkindji cultural claims of a special cultural relationship to the River since they consider Barkindji claims to culture as spurious or imagined. At a basic level this is a denial of Barkindji claims to a deep rooted connection with the land as its first peoples (Tuan 1980).

There is a widely held public perception that the 'identity and rights of contemporary Aboriginal people depend on them being recognized as in some sense "the same" as those who occupied the country when the settlers arrived' (Beckett 1988: 6). Because Aborigines in the south-east of Australia were colonized early and it is the view of many whites in Wilcannia, and indeed of the wider Australian public, that only those Aborigines living in 'remote' areas of the North and Centre of the continent are 'real Aborigines' or retain Aboriginal culture of any worth or 'authenticity'. According to the popular view, living a 'traditional' hunter-gatherer way of life is what 'real Aborigines' do. Those Aborigines who share white space and place, such as those in Wilcannia, are seen by most local whites to have 'lost their culture'; more pertinently, they are thought to have lost any cultural ways that are acceptable to whites and which whites value. This sees Wilcannia Aborigines perceived by the wider public as having lost their 'traditional' spiritual connection to place and to land. It is this 'traditional' connection which is valorized by, and lives in the imagination of many settler Australians (Lattas 1990). It is one which whites accord other Australian Aboriginal groups living in the more remote Centre and North of the continent who are seen to live 'traditional' lives. According to the popular view, and indeed the historic view of 'salvage' anthropology, the more proximal Aboriginal people are to white settlers in time and space, the greater the perceived reduction of Aboriginal culture. This view implicitly assumes that Aboriginal culture is something that can be lost but cannot be added to (Byrne 1996). According to these notions one can lose one's cultural sense of place. Whites, in perceiving that only 'tribal' Aborigines have Aboriginal culture, are operating on the assumption that Wilcannia Aborigines are in many ways no different to whites. However, although most whites do not accord Aboriginal culture to Barkindji people, and recognize black and white similarities of daily life, they do accord them Aboriginal *difference* in ways that work against the cultural health and wellbeing of Barkindji people.

Local whites implicitly recognize some notion of Aboriginal cultural difference in the anthropological sense of 'a people's way of life', but they simultaneously assert a lack of 'Aboriginal culture'; they say, 'They don't have any culture' and 'What culture?' These judgements are based on the dominant culture notion that authentic Aboriginal culture is that which bears little or no resemblance to white ways. Aboriginal culture here can be differentiated from Aboriginality, which is accorded, and is for most whites in Wilcannia, associated with negative traits and behaviours. The high incidence of alcohol abuse, domestic and more general intra-group violence, high unemployment rates, the perceived tendency towards immediate gratification and a lack of parental and financial responsibility are seen by many whites to be immediately recognizable aspects of being Aboriginal in Wilcannia. So while most whites in Wilcannia deny Aboriginal people a positively-valued Aboriginal culture, they nevertheless attribute Aboriginal people with Aboriginality – although of an undesirable kind (Gibson 2010). That is, they do not deny Aboriginal heritage and difference of a kind; but this is one which operates against the forms of recognition that Honneth considers to be crucial aspects of identity formation: those of "self-confidence, self-respect and self-esteem" and the opportunity for the shaping of an "attitude towards oneself" (1995: xi, xiii): all of which one can reasonably argue are necessary for one's sense of wellbeing and place in the world.

The capacity of Aboriginal people in Wilcannia to self-determine, as with their urban counterparts:

> will continue to be elusive while their history and their culture is mediated by others, and the evolution of Aboriginal Identity takes place, not in an autonomous fashion, but as a response to the construction of knowledge by members of white society. (Jordan 1988: 128)

Because Aboriginal people are called to construct their identities in response to the knowledge and social sanctioning of white society, and because Aboriginal culture has mostly been disallowed for those in Australia's south-east, self-determination is not only difficult but has become a more conscious, responsive and therefore increasingly reflexive task (Correy 2006, Gibson 2008). A blindness to the fluid and living nature of culture by the wider public, who preference more essentialized hermetic structures of Aboriginal culture has seen ideas of Aboriginal culture become, in effect, fixed as at the time of colonization. This is reflected in the Aboriginal Land Rights (Northern Territory) Act (1976), legislation which can confer rights to land and resources and which 'demand[s] evidence of the continuity of traditional beliefs, practices and dispositions as the condition of cultural recognition' (Povinelli 2002: 3). Because many of the Aboriginal people of the south-east have been dispossessed of their land, they have been unable to demonstrate continuity of traditional practices and dispositions in relation to their land and the places that they hold dear in the space of the imagination and in story. Dispossession of their rights to determine their cultural identity is a direct consequence. The ostensible lack of overt, and what are readily understood to be, Aboriginal cultural practices by Barkindji has seen local and other whites view Aboriginal culture in the south-east in terms of *lack*.

The Dreaming is an 'Everywhen'

In order to understand the effects that the white denial of Barkindji culture has on the wellbeing of Barkindji people, knowledge of the Aboriginal concepts of 'country' and the 'Dreamtime' or 'Dreaming' is necessary. Barkindji country is thought to have been created in the Dreaming, a complex concept understood by Aboriginal people in various ways best captured in anthropology by Stanner (1979):

> The Dreaming conjures up the notion of a sacred, heroic time of the indefinitely remote past; such a time is also, in a sense, still part of the present. One cannot "fix" The Dreaming *in* time: it was, and is, an *everywhen*. (Stanner 1979: 24; my italics; see also Elkin 1968)

During the period of the Dreaming, ancestral beings, both human and animal, moved across the land creating its topographical features, sometimes meeting, sometimes fighting, some coming to rest as features such as hills and rocks and some going into the ground where their essence continues in active relationship with 'country' (Munn 1973). The older western term and belief of *genius loci*, that is, spirit of place, whereby places were thought to be guarded by spirits and where people paid homage to these spirits through ceremony and ritual continues to resonate with Barkindji ideas of place and 'country' (DeMiglio and Williams 2008: 16). Known also as the Law, the Dreaming is when the rules, practices and moral framework for life was laid down to be followed; 'a kind of narrative of things that once happened; a charter of things that still happen; and a kind of *logos* or principle of order transcending everything significant for Aboriginal man' (Stanner 1979: 24). The country of one's Dreaming is here both a 'common noun and a proper noun'; it is not simply 'imagined or represented, it is lived in and lived with' (Rose 1996: 7). Country, in this worldview, knows, feels, hears, smells, acts and communicates; it lives in and through time and 'has its own life, its own imperatives, of which humans are only one aspect' (Rose 1996: 10). The Dreaming and the River are interconnected in important ways.

The Darling River which runs through Wilcannia was, according to many of the Aboriginal inhabitants, created by Ngatji, the ancestral Rainbow Serpent. While Barkindji state that there are male and female Ngatji as this is necessary for reproduction, most refer to Ngatji as male, as will I. Ngatji is said to have moved across country carving out the Rivers and deep waterholes on Barkindji country. Much of the earlier writing regarding Rainbow Serpents concerns the distribution of these myths around Australia, their purpose, their association with beliefs, customs and ritual, particularly that associated with rain-making and water, and the relationship between the Rainbow Serpent and minerals such as quartz and opal (Buchler and Maddock 1978). Although it is certainly the case that Barkindji people in Wilcannia no longer live as their ancestors did and no longer have a system of knowledge or ritual in the way of their ancestors, there is nevertheless a strong general belief in Ngatji's presence and his contemporary capacity to intervene in human/nature relations. When the Department of Land and Water Conservation (DLWC) began water exploration and sank a bore in Wilcannia in late 2003 in an attempt to secure drinking water for the town, Barkindji beliefs in Ngatji were fore-grounded. Described by Barkindji as both 'good and bad', Ngatji is an ambivalent figure who is said to live in deep waterholes, sometimes swimming in underground water tunnels between specified points. Several of the older Barkindji people were worried that the DLWC would find Ngatji's underground water tunnel and that this might 'stir Ngatji up'.

One of my Aboriginal friends aged in his early sixties said, 'We believe the Ngatji should be left alone because if ... anyone muck with it, it's gonna hurt the people; maybe because we are the protectors'. I asked, 'Even if it was a white ... a white person that mucked with it ... would it hurt *Wiimpatjas*?'[2]

> We – well, that's what we believe – that the Ngatji is there to give us our fish, he looks after us with our water, you know, the soaks in the river ... an' ... we believe that if ... anyone muck with it ... it ... you shouldn't muck with it ... it should be just left alone.

Many Barkindji people consider that the white man has completely 'mucked up' the River partly as a result of the cotton farming and irrigation 'up river'. As Badger Bates a Barkindji man and well known cultural authority explains, 'They're drainin' our rivers out and it's breakin' our heart'. Whites are thoroughly implicated in affecting the place and wellbeing of Ngatji and the River and, in turn, Barkindji people in terms of whites being seen to drain the rivers, yet the onus of responsibility for these actions in terms of accounting to Ngatji appears to rest with

---

2 *Wiimpatjas* is a term sometimes used by Barkindji when referring to themselves.

Barkindji. Here, the Barkindji sense of wellbeing is outside of Barkindji control linked at it is to mainstream policies and programmes of various government and non-government agencies and bureaucracies.

While I was living in Wilcannia it was experiencing the worst drought in over one hundred years. No water ran over the small rock weir in town and river red gums began growing on the river-bed. Both Aboriginal and non-Aboriginal people agreed that the River was, and indeed currently remains, in crisis but how this came about and how this might be remedied is differently *thought.* For the most part the crisis of the River is, for whites, one of economic and environmental cause and concern in a Western worldview where the environment is exterior, perhaps capricious, but where humans 'may nevertheless seek mastery over it' (Croll and Parkin 1992). On the other hand the crisis of the River is entirely bound up with the crisis of the people in a network or relations which does not make a human/ environment distinction in these ways. The interconnectedness of the natural world, of animals and of humans was an expressed ideal for most Barkindji. This is not to say that Barkindji cosmology and western scientific ways of knowing are incompatible as many Barkindji accommodate and encapsulate these within their view(s) of the world.

Because culture is always fluid and in flux, belief and knowledge systems of Barkindji people – as with non-Aboriginal people – change over time. As a result, place and relationship with place are subject to multiple imaginings, manifestations and ways of knowing. I was talking to a younger Barkindji friend in his late twenties who, as with many of his peers, had received a Western education. I asked him what size Ngatji was. My friend said that he was not really sure because he was not sure if he had actually seen Ngatji, although he thought he may have. He elaborated, describing how he and some friends had been walking across the Wilcannia Bridge at night when they saw something moving fast in the water under the bridge. Although they were not sure it was Ngatji, they deduced by a process of elimination that it may have been. They considered each of the species they knew and associated with the river in terms of the behaviour of the River and the shapes and shadows that they could see; it was not cod, cat-fish, and so on. My young friend added that his Nan had seen Ngatji and relayed his Nan's description of Ngatji's size (indicating a length of about three metres) and having the girth of a 'Sunshine Milk tin'. I asked whether Ngatji was thought to have physically carved out the River and how his[3] size might relate to that. My friend said he had never really thought about it 'in that way', he said he just thought that Ngatji existed. In terms of Ngatji's physically carving out the river, he said that Ngatji was 'an Aboriginal explanation for the creation of the land'. He thought a bit more and then alluded to the time of the dinosaurs: the 'original' Ngatji may be like the 'mega fauna'.

---

3   There are both male and female Ngatji but, in most discussion Ngatji is invariably referred to as "he".

When I asked the same question of a much older Barkindji man, he replied by asking me this question: 'How does a small creek begin?' In response I ventured, 'Is it erosion?' He then asked me whether I had ever seen a track made by sheep[4] and if I knew what happened to such a track after rain. He went on to say that what was a small creek millions of years ago might now be a river. He suggested that Ngatji may have created the creek, the beginnings of the River, and over time 'nature' had widened it.

Although my line of questioning might be considered inappropriate (that is, asking a cause and effect question of what is considered to be a 'mythical' being) these two examples tell us is that the scientific, the evolutionary and the mythic are not necessarily at odds with one another. In fact, these perspectives are readily accommodated within the epistemology and ontology of some Barkindji people in ways that, importantly, do not reduce the belief in Ngatji's presence or capacity to act in the world. Yet, because Barkindji can and do accommodate evolutionary views within their ontology, this is viewed by many whites as yet another indication that Barkindji claims to a particular and special relationship with the River are, if not spurious, certainly inauthentic.

The physical and spiritual interconnectedness of everything in nature remains cogent through the teachings of certain Barkindji Elders who operate in the world accordingly. Ngatji's dwelling place is subject to the actions of non-Aboriginal others, and is largely outside of Barkindji control, making it difficult for Barkindji to ensure Ngatji is placated and in balance with Barkindji and all other living things. Responsibility without control creates problems for cultural and spiritual wellbeing.

The following ethnographic example highlights the disparity in some Aboriginal and non-Aboriginal ways of thinking about the nature and supernature of the world and how this links to different cultural senses of place and indeed wellbeing. During fieldwork, I attended many meetings of the Wilcannia Community Working Party (WCWP).[5] During the height of the drought in 2003, and despite severe water restrictions imposed by the Central Darling Shire Council (the Shire), the town water supply was near exhaustion. The drinking water, though filtered, was unpalatable and emitted a foul algal odour; salt levels were unacceptably high. At a WCWP meeting, the Shire's General Manager announced that the River was filling up below the weir despite a lack of flow over the weir for several months. The General Manager said that Shire staff was trying to locate the source of this

---

4   Barkindji sometimes refer to sheep as 'ground lice', since they damage their country.

5   The WCWP is a group of people chaired and governed by local Aboriginal people with invited non-Aboriginal membership across the various agencies and institutions which service the town; included are representatives from local schools, the police, government human services departments, the local women's safety refuge and the Aboriginal Land Council. A Memorandum of Understanding with the NSW government recognizes the WCWP as the appropriate body for consultation on matters affecting the local Aboriginal community.

water but that it was 'a mystery'. After the meeting an Aboriginal man explained to me that the mystery was indeed no mystery, and neither was it a mystery to several other Barkindji people I spoke with about this matter. It was interpreted as Ngatji having supplied the water. Two women in their fifties commented that even prior to the water appearing, they had known there was something happening with Ngatji because of the behaviour of the water under the Wilcannia Bridge. Others said, 'Ngatji knew we needed the water' and 'We knew he was still here'. For many Barkindji people, the additional water was confirmation of what they already knew: it was a 'sign' of Ngatji's ongoing presence and his response to their need.

A few days after the meeting, the Shire explained that the source of the water had been found: the release valve from the town's water filtration plant had been left open and the pump had been left running. The overflow from the ongoing pumping was going into the town's drainage system and then flowing into the river below the weir. In white terms, the previously unexplained source of water was now appropriately accounted for and was therefore no longer a 'mystery'. This difference in accounting for the water below the weir may be viewed as being a difference of opinion or a clash of cultural beliefs which does little harm, yet when whites through their words, looks and actions openly disregard or pay lip-service to Aboriginal cultural beliefs, as they often do, this is damaging as it is dismissive of the reciprocal responsibility that many Barkindji people believe exists between themselves and Ngatji. If one cannot perform one's cultural obligations in relation to place then reciprocal affects on wellbeing can be assumed.

When the River did finally receive water following rain further upstream, Karen Riley, a Barkindji woman, showed me a letter she had written and was preparing to send on behalf of the Wilcannia community to the main local newspaper. In it she wrote that 'when the River started flowing again we could all feel a sense of calm ... The unique feeling that it gives us when the river runs, is felt by all in our community'. She explained that when the water stopped flowing, 'we stopped camping out on the river and it seemed we started losing our sense of family. Our children became wayward and it appeared as if our people started losing control of their lives'. As this letter shows, physical, mental and social health is linked to the health of the River and a healthy cultural way of life. A 'sick' river means that people cannot fish any more, impacting social relations and the ability to supplement diet and income. Local whites do not understand the connection between the sick River and its consequences as described by Karen and others; what whites see to be Barkindji problems of alcohol, violence and family dysfunction are attributed, not to a dry river-bed and cultural thirst but to social laziness and irresponsibility.

The figure of Ngatji does not always respond or act in ways that are beneficial to Barkindji or their needs. Much of the information about Ngatji is relayed to Barkindji children as warnings and descriptions of dire consequences of unheeded warnings. In recounts largely by older Barkindji women, drowning – for those children and adults whose lives end this way, is predestined. When a small boy drowned, Barkindji people said that 'the river called the child down there' and that 'it was his time'. This was not the view of many whites, who cited poor parental supervision, and who viewed Barkindji explanations as superstitious or deluded. At the same time Barkindji warnings about Ngatji's will and power to drown are heavily invested in, which suggests that drowning may have the potential to be avoided thereby speaking to a level of free will. Ngatji's ambivalent nature is however clear. One man said Ngatji is 'mainly good, bad when you break the Law, but you always get a warning'.

Irrespective of the extent or form of belief in Ngatji there is a general adherence to everyday protective sanctions in relation to place and staying well. One must 'use a low voice near Ngatji's *mingga* (waterhole)' to avoid disturbing him. I was told several times when near such a waterhole that we must not speak the name. One Barkindji man in his fifties told me of an incident that occurred when he was on the riverbank with his two sons and his partner. His son was swimming in the 'right' place but he must have been a little too near Ngatji's hole and the man saw 'the water comin' up the River'. He yelled for his boy to get out and said he told Ngatji: 'We don't mean any harm', which seemed to pacify Ngatji, who went away. The man considers, 'Old people can control Ngatji if you talk to it properly'. Other signs of Ngatji's presence, I was told by this older man, are when 'whirly winds form in the water' and when 'water squirts up'. I asked him why Ngatji should not be spoken about near his *mingga* and I was told:

> If you talk too much it'd come up an' get angry. Because if we talk about Ngatji too much you know, then white people an' other people, they can come along and start muckin' around disrespectin'. An' if you disrepectin' Ngatji too much he'll come and he'll drown someone.

The River here is more than a coherent symbol of cultural value; it is closely linked to physical safety, together with cultural health and wellbeing. It is to Ngatji that many Barkindji people look to provide them with fish within a relationship of reciprocity. People 'sing out for fish'. One Elder with whom I fished would shout out '*Parntu!*' [cod] before casting his line. He said that singing out in this way 'mean I want fish' and want Ngatji 'to push fish towards me. If I want cod I shout *Parntu!*, shout *Kuuyu!*, any kind of fish. Even today Ngatji still protect us, if we look after him – you know?' In understanding how place and wellbeing operate in relation to Ngatji it is important to recognize that Ngatji is not confined to the physical or spatial River and shows his pleasure when he 'blows a rainbow' in the sky after rain:

> It means for them to blow the rainbow that they happy, they know the people who they're protectin' and who they're for. But we happy cos we're getting food out there. Like plants come up, also the animals what we eat. The rainbow means that you're gonna get a lot.

The relationship of Ngatji with humans, animals, plants and the land itself, as with other ancestral beings, is one of interconnectedness. The way in which the water of the River is managed however, places strain on this relationship, further reducing the potential for Barkindji social, economic and cultural wellbeing. Currently, the share of water that reaches Wilcannia from its headwaters at the border between the states of Queensland and NSW is politically contested. The Darling River is part of the Murray–Darling Basin system which covers 14 per cent of the Australian continent and supports over 70 per cent of the nation's irrigated crops and pastureland (Crabb 1997). The Basin is critically important, politically and economically for the dominant culture. The Basin authority together with other water related agencies is required by legislation to consult with Barkindji people over the management of the River as recognition of their cultural interests. This has seen the River become a tangible focus and symbol of 'Barkindji culture'. Yet, this is against a backdrop of a perception by many of the small white population of Wilcannia that Barkindji culture has all but died out.

The notion of The River is a central and recognized theme of Barkindji culture by Aboriginal and non-Aboriginal people in Wilcannia is, however, problematic. On the one hand we might understand and consider that Barkindji expressions which appear to purposefully link identity to the River are a verbal imperative. They are perhaps crucial and necessary as a political moment for the way that they appear to somehow crystallize some notion of Barkindji culture and identity where the presence or 'authenticity' of Barkindji culture is itself contested by many whites.

On the other hand, there is a sense that the expression of 'being the River' is, for many whites, a symbolic 'artifact' that has come in some ways to 'stand for', or at least 'stand in for' Barkindji people. Where official consultation with Barkindji over the management of water is required, the promotion of a 'taken for granted' oversimplification of Barkindji identity as 'being river people', as 'belonging to the river, and 'being the river', glosses over the ontological complexities of living people with many different experiences and ideas about what it is to be Barkindji. Being 'people of the River' has become a catch-cry, which entombs living culture and the multiple interests and identities of Barkindji people in an over-simplistic and limiting discourse. An example being when the concomitant and inferred spatial and physical bounded-ness of belonging to the river is played out, and fed into policy and programmes which separate and categorize the control and management of riverine environments. Here, aspects of being Barka and being Barkindji – aspects which for Barkindji people are connected, inter-relational and experiential – may become fragmented and compartmentalized within agency structures as either 'natural resource management' or 'cultural heritage'. That is, there is a separation of the supposed 'environmental' and the 'cultural': Dreaming sites, particular fish species, fish ladders, weirs and so on are treated as separate aspects and locations which fragment the holistic Barkindji sense of place with associated consequences of cultural health and wellbeing.

The nature of Western management systems of categorization and sub-categorization continue to divide, separate and compartmentalize what most Barkindji people do not separate. The practice of categorization and taxonomy sees natural resource management further broken down into subcategories of attention such as, land and vegetation, biodiversity, and salinity each having their own 'action' plans. When Barkindji people are consulted on matters relating to the River, all too often this is attended to within either an ecological framework or a cultural heritage category. Moreover, natural resource management, although emphasizing a continuation with Aboriginal practices of the past, is concerned with the present, the future, and the 'natural' as opposed to the 'supernatural' environment. For the most part, natural resource management seeks to document Aboriginal knowledge of ecology in the belief that it will enhance sustainable development. Aboriginal Cultural Heritage, on the other hand, concerns the protection and maintenance of visible material culture, landscape forms and sites associated with stories and myths from the 'remote past'. It seeks to preserve a past culture, rather than to encourage or recognize contemporary forms of living cultural practice.

One young Barkindji man of about 30-years-old told me:

> This river, Darlin' River – they call The Darlin', an' us Barkindji people call it 'Barka' ... is very ... very important ... because it's part of my identity, it's who I am. When I was growin' up I was bein' told ... 'Barkindji people river people' ... So I was always told, that was where we come from, and ... that we're part of it. Ahm ... I was lucky enough, like a lot of people around here who grew up with the old people, who knew their old people ... an' growin' up, we learnt that they really respected their country. My best moments in my life, I remember learnin' my land from my old people an' stuff ... An' learnin' about family values and everything like that there ... cos I'm the river.

Not only does this man draw his identity from the Barka, he is the Barka. Part of being the Barka and being Barkindji involves not only respect for the River but for 'country' and for family values. There is a holism here which is absent in the strategies, policies and programmes of government and non-government agencies despite their often good intentions to the contrary.

Conclusion

Both Aboriginal people and non-Aboriginal people in Wilcannia have their own ideas of what constitutes belonging, wellbeing and a sense of place. As a consequence, overt verbal identity markers such as 'belonging to the river', 'being River people' and 'being the river' are a double-edged sword for Barkindji. Most whites seem prepared to accommodate these Barkindji expressions and relational claims as tokens of a symbolic Barkindji association. This said, the holistic nature of cultural identity whereby the River is part of a lived ontology and cosmology is not recognized or accommodated by most whites. Most whites have difficulty in reconciling the Barkindji way of living with what they imagine or consider 'Aboriginal culture' to be. When Barkindji elucidate what it is to 'be Barka' in all its cultural fullness which includes behaviours of which whites disapprove then Barkindji assertions of culture are met with white cynicism and scepticism. This has seen Barkindji identity and cultural knowledge fractured and made malleable in ways that are detrimental to cultural knowledge, cultural identity and therefore cultural wellbeing.

In sum, place and wellbeing, their conceptualization and experience result from relationships between people places and things (DeMiglio and Williams 2008). Cultural conditioning and worldviews intervene in these relationships in both conscious and unconscious ways which ultimately favour the dominant culture to the detriment of Barkindji senses of place and wellbeing. As we have seen, Aboriginal and non-Aboriginal cultural drivers and constructions of health, living conditions and quality of life in relation to the River, and what it means to be in the world, are differently perceived, imagined and lived, as indeed are experiences and imaginings of place.

References

Australian Bureau of Statistics. 2001. *Measuring Wellbeing: Frameworks for Australian Social Statistics*. Canberra: Australian Bureau of Statistics.

Beckett, J. 1988. The past in the present, the present in the past: The construction of Aboriginality in *Past and Present: The Construction of Aboriginality*, edited by J. Beckett. Canberra: Aboriginal Studies Press, 191-217.

Bloch, M. 1992. What goes without saying, in *Conceptualising Society*, edited by A. Kuper. London: Routledge, 127-46.

Buchler, I.E. and Maddock, K. 1978 (eds). *The Rainbow Serpent: A Chromatic Piece*. The Hague, Paris: Mouton Publishers.

Byrne, D. 1996. Deep Nation: Australia's Acquisition of an Indigenous Past. *Aboriginal History*, 20, 82-107.

Correy, S. 2006. The reconstruction of Aboriginal sociality through the identification of traditional owners in New South Wales. *The Australian Journal of Anthropology*, 17(3), 336-47.

Crabb, P. 1997. *Murray Darling Basin Resources*. Canberra: Murray Darling Basin Commission.

Croll, E. and Parkin, D. 1992. *Bush Base, Forest Farm: Culture, Environment and Development*. London, Routledge.

DeMiglio, L. and Williams, A. 2008. A Sense of Place a Sense of Wellbeing, in *Sense of Place, Health and Quality of Life*, edited by J. Eyles and A. Williams. Aldershot: Ashgate, 15-30.

Elkin, A.P. 1969. Elements of Australian Aboriginal Philosophy. *Oceania*, 40(2), 85-98.

Gibson, L. 2008. Art, culture and ambiguity in Wilcannia New South Wales. *The Australian Journal of Anthropology*, 19(3), 294-313.

Gibson, L. 2010. Making a life: Getting ahead and getting a living in Aboriginal New South Wales. *Oceania*, 80(2), 143-60.

Jordan, D. 1988. Aboriginal identity: Uses of the past, problems for the future, in *Past and Present: The Construction of Aboriginality*, edited by J. Beckett. Canberra: Aboriginal Studies Press, 109-30.

Lattas, A. 1990. Aborigines and Contemporary Australian Nationalism: Primordiality and the Cultural Politics of Otherness, in *Writing Australian Culture: Text, Society and Cultural Identity*, edited by J. Marcus. Adelaide: University of Adelaide, 50-69.

Munn, N. 1973. *Walbiri Iconography: Graphic Representation and Cultural Symbolism in a Central Australian Society.* Chicago, IL: Cornell University Press.

Norberg-Schulz, C. 1969. Meaning in Architecture, in *Meaning in Architecture*, edited by C. Jencks and G. Baird. New York: George Braziller.

Povinelli, E. 2002. *The Cunning of Recognition – Indigenous Alterities and the Making of Australian Multiculturalism.* Durham, NC: Duke University Press.

Relph, E. 1976. *Place and Placelessness.* London: Pion.

Rose, D. 1996. *Nourishing Terrains: Australian Aboriginal Views of Landscape and Wilderness.* Canberra: Australian Heritage Commission.

Radcliffe-Brown, A.B. 1926. The rainbow-serpent myth of Australia. *Journal of the Anthropological Institute of Great Britain and Ireland*, 56, 19-25.

Stanner, W.E.H. 1979. *White Man Got No Dreaming: Essays 1938-1973.* Canberra: Australian National University Press.

Tuan, Y.F. 1980. Rootedness versus sense of place. *Landscape*, 24, 3-8.

Tuan, Y.F. 2005. *Space and Place: The Perspective of Experience.* Minneapolis, MN: University of Minnesota Press.

Chapter 14

The New Therapeutic Spaces of the Spa

Jo Little

This chapter examines the contemporary spa as a therapeutic space. In so doing it contributes to the expanding literature within geography on the relationship between place and wellbeing. In highlighting the arguably rather neglected spaces and practices of the spa, the chapter provides empirical detail of an original therapeutic environment. It also moves beyond and contests ideas of wellness and the healthy body that lie at the centre of work on therapeutic landscapes by exploring the relationship between wellness and bodily regulation. The chapter thus offers an interpretation of spa spaces and practices as both indulgent and disciplinary focusing on the ways in which the idea of the healthy body is highly dependent on the (gendered) assumptions surrounding bodily size, shape and appearance.

The chapter is informed by three main areas of geographic literature and debate. First is work on therapeutic landscapes and ideas surrounding the co-construction of health and place. Here I develop recent calls to extend work on the health-giving qualities of particular landscapes and environments (see Curtis et al. 2007, Williams 2007) in a consideration of the performative relationship between the self and place. This involves examining how therapeutic spaces are experienced, valued and understood by the individual. Such an approach has encouraged the development of ideas of wellbeing and has also helped to foreground the ordinary spaces of healing and the more routine and day-to-day wellbeing practices and encounters (Milligan 2007).

The second substantial area of work on which the chapter draws is that on gender, the body and identity. This research is used here to draw attention to the role and importance of spa visits as part of the embodied construction and performance of femininity. Recent literature on embodiment is used to show how particular constructions of the body have been normalized within dominant gender subjectivities and how these constructions are central to contemporary ideas of wellness and fitness. This is linked to the third set of literatures that inform the chapter, namely work on the regulation of the body. The chapter explores the ways in which certain therapeutic practices, while emphasizing fitness and wellbeing, act in a disciplinary way to control of the body and ensure compliance with the bodily stereotypes associated with particular gendered identities. After situating the chapter within a review of these various literatures, I go on to explore the issues in discussion of original research conducted in South West England. As explained in a later section, this study included a questionnaire with spa visitors as well as interviews with the managers of two spas.

Wellbeing and the Spa: Key Themes and Debates in Geography

Therapeutic Landscapes and the Spa

The notion of the therapeutic landscape has commonly been employed by geographers interested in the ways in which space and place become implicated in ideas of health and healing. There is now an extensive critical literature exploring the characteristics of those environments perceived to be in some way 'healthy', and the qualities such as remoteness, peacefulness, solitude, purity and beauty associated with the therapeutic capacities of such landscapes (Kearns and Gesler 1998, Williams 1999). In a review of the development of more recent approaches to the study of the relationship between environment and health, Curtis et al. (2007) suggest that the concept of therapeutic spaces stems from three main lines of thought. First, they note, is work from cultural ecology and environmental psychology from which ideas about the healing properties of nature and the relevance of building design has emerged. Second, work from a structuralist tradition which emphasized issues of social interaction, power and marginalization within health settings and third, humanist work on ideas about disease and the values of experience and feeling in particular spaces.

These different strands of research reflect a development of the therapeutic landscape concept from more static ideas about the health-giving value of certain environments to examine the relational dimension of the self-landscape encounter and consider the ways in which people interact with the landscape through a complex set of transactions. Such approaches have encouraged a more holistic consideration of health and wellbeing in people's encounters with place and recognition of the 'broader web of socio-natural relations within which the individual is imbricated' (Conradson 2005: 338). The therapeutic landscape concept, then, has framed work on the design and use of particular spaces of care and on the relationship between health care spaces and identity. It has also incorporated and extended ideas on the social construction of illness in the context of different sites and spaces (Parr 2007). In line with all these developments has been recent work on the affective body in relation to therapeutic spaces and to the centrality of emotions in understanding everyday geographies of health and wellbeing (Lea 2008). Such work has drawn on theoretical ideas surrounding the 'care of the self' (as discussed below) and has recognized the ways in which spiritual connections with place are a part of the broader association between wellbeing and place.

In addition to these conceptual developments, work on therapeutic landscapes by geographers has also encouraged a shift away from iconic and formal sites of healing to include more localized and ordinary spaces of care and wellbeing as well as the more everyday practices and routines associated with health and therapy. These moves have extended the idea of therapeutic landscapes to include places of relaxation, wellbeing and fitness. Such developments provide an important context for the study of the contemporary spa as a therapeutic environment and as a space for healing practices and wellbeing.

Spas as therapeutic spaces have largely been studied in their historical role as medicalized spaces although some work has focused on the social and power relations (including gender) operating within the spa as well as on the meanings, practices and embodied relations associated with them. In his work on Bath spa, Gesler (1998), for example, explores various themes that reflect, he suggests, the historical importance of spas as sites of healing and social interaction. He discusses the relationship between the physical and symbolic values attached to spas and their social characteristics and organization.

Other writers have also sought to locate the importance of spas and other 'healing spaces' within broader frameworks of social organization and ideology; Cayleff (1988: 85), for example, considers the 'water-cure' movement of the nineteenth century and the ways in which spaces of hydro therapy reflected 'a reformist and even emancipationist outlook' toward class and gender attracted those 'seeking order, self determination and a more empowering view of health and life's meaning'. As Foley (2010: 10) notes in the introduction to his study of therapeutic landscapes in Ireland, '(h)ealing waters are the product of a range of cultural narratives and performances from the religio-magical to the pseudo-scientific, wherein healing is expressed through words and feelings as much as in physiological outcomes'.

In terms of this wider relationship between the spa and social relations, work has demonstrated a shift in the nineteenth century away from an elite clientele to more general appeal. As part of this shift, the spa, like other water-based healing activities such as sea bathing, became more associated with leisure than with health and, as a result, with 'lower' as opposed to 'upper' class identities (Foley 2010). Studies of the popular appeal of spas at this time have drawn on ideas of the carnivalesque and in particular the work of Bakhtin (1984) in discussion of performance and identity within these liminal spaces. Such studies have also pointed to potential conflicts between health and sociability recognizing the ways in which some spas and other therapeutic spaces became associated with excess and over-consumption.

As noted above, there has been relatively little attention given to the contemporary spa as a therapeutic environment. This chapter, therefore, seeks to highlight the curative functions of the spa and also, like studies of other therapeutic landscapes, to explore the embodied practices and performances through which the spa as a space of wellbeing is constituted. In so doing the chapter emphasizes the broad functions of the contemporary spa and the ways in which spas foreground relaxation and pampering as part of the healing process. It confronts the sense in which the journey, as Foley (2010: 117) puts it, 'from stress to de-stress' is integral to the spa's role in supporting wellbeing. The study of the contemporary spa also shows how embodied practices of therapy and wellbeing can be seen as both therapeutic and also disciplinary. To some extent this picks up on the tension identified earlier in the study of 'healing waters' between 'health' and 'pleasure'. Questions are raised concerning the luxury and beauty therapies of the modern spa with their emphasis on consumption and celebration, and the ways in which they unsettle the boundaries between wellbeing and indulgence, health and discipline.

Spas, it is argued, are in the UK at least, becoming a common and accepted part of regular beauty and health regimes. They are a part of what writers such as Straughan (2010) have referred to as 'the body industry' – places in which the fit and healthy body is seen as a physical and emotional project. At the same time they are exclusionary places, promoting a particular attitude towards the body and gender and while most would present a highly accommodating and welcoming approach to the body, they nevertheless are spaces that discipline the body through moralistic messages about fitness, shape and bodily appearance.

Spas, Gender Identity and Bodily Discipline

The notion of the spa as both therapeutic and disciplinary positions this study within a wide range of geographical literatures. Understanding the role of the spa in exploring contemporary notions of wellness and the body draws on recent work on gender identity and, in particular, the construction of femininity. Spa use, as confirmed below, has been acknowledged to be strongly gendered with the majority of visits being undertaken by women and, as such, needs to be seen in the context of research and writing not only on the gendered nature of leisure but also on the relationship between leisure/wellbeing and embodied subjectivity. As will be argued in the empirical section of the paper, spa visits and treatments highlight the importance of relaxation and pampering to women's leisure and to their gendered identities. This importance, it may be argued, relates to pressures to achieve and maintain a particular kind of body and also to changing power relations within the workplace and their influence on feminine identities.

Feminist researchers have long been interested in ideas of the appropriate body and have explored the social, cultural and moral pressures that have shaped expectations around the acceptable and unacceptable body. In exploring the ways in which the gendered body is controlled and disciplined, they have drawn extensively on the work of Foucault. Bodily regulation is shown, within such work, to be 'part and parcel of a more general process of surveillance and control in modernity' (Holliday and Thompson 2001: 117) that operates through social practices in different ways in different places. Foucault's concept of biopower has been seen as particularly relevant to the critical (post-essentialist) examination by geographers of the use and presentation of bodies and of the diverse authorities concerned with their health and utility. Foucauldian perspectives view social practices (particularly those concerned with medicine and psychiatry) as practices of control that produce certain kinds of bodies and psyches. Thus 'individual identities are shaped through the imposition of normative labels and categories and bodies are made docile as they are socialised' (Pitts-Taylor 2007: 24) and policed by the self through internalization of external 'panoptic' measures.

The role of the beauty industry has been highlighted as increasingly important in setting and achieving the characteristics of a desirable and acceptable body (see Petersen 2007) and to conflicting notions of power and exploitation in women's attempts to conform to accepted bodily ideals of size and appearance. Links have been made to the gendering of the healthy body in contemporary medical discourses (Black and Sharma 2001, Moore 2010) suggesting that women carry a greater responsibility for monitoring family wellbeing and for ensuring that the wider conditions for health and fitness are met (particularly in relation to diet and lifestyle). Black (2002, 2004) brings these perspectives together in her work on pampering and the feminine body. She argues that family and workplace pressures on women in which they are expected to provide emotional and productive labour have prompted a growth in the need for defined times and spaces of relaxation. Taking care of the body has thus become as much about rest, relaxation and recovery as it has about the treatment of disease and illness.

Straughan (2010) also explores the relationship between the medical and the social in the practices and spaces of the beauty salon. Like Black, Straughan sees the emergence of the contemporary salon in the context of the development of beautification as a science rather than an art and a corresponding shift in ideas about the modification of women's bodies as personal, public and political. Like other geographers (see Lea 2009, Parr 2002, Philo 2000), Straughan (2010: 650) draws on Foucault's concept of the 'care of the self' arguing that his ideas on the 'aesthetics of existence' provide a useful framework for unpacking the 'technologies, treatments and practical engagements' of the beauty clinic as important disciplinary strategies which incorporate medical and emotional aspects of wellbeing. Although criticized for rendering the individual docile and lacking in agency, the 'care of the self' is seen as particularly helpful in exploring the ways in which the body is fragmented by the scientific focus of beauty and wellbeing strategies.

The Spa as a Therapeutic Landscape

The Research

The data used in the remainder of the chapter are drawn from ongoing research on spas and spa clients in the UK. This research incorporates interview data from in-depth interviews conducted in two spas in Southern England, together with questionnaire data and textual material. Both spas studied were part of bigger leisure or holiday complexes. The larger, the Gloucestershire spa, was part of a business involving a hotel, restaurants and a bar/pub while the Devon spa was simply an addition to a small group of holiday cottages. While the Gloucestershire spa was more aggressive in terms of growth and customer base, both spas were expanding businesses and both were clearly taking advantage of an expansion in demand.

Basic quantitative data on spa use and the kinds of treatments and experiences they offered were gained from a questionnaire of spa visitors conducted over a short period in one of the spas. This was a self-completed questionnaire which guests were invited to fill out during their visit. While the questionnaire did not provide much detail in relation to the views and reactions of the spa users it was useful in building up a picture of the nature of the spa's clientele, the reasons for their visit (particularly their use of the additional spa treatments) and how the spa made them feel. In total 50 questionnaires were returned.

The in-depth interviews were carried out with the mangers/owners of the spas – three interviews of about 2 hours each were undertaken and fully transcribed. The interviews were semi-structured and explored issues concerning the spa businesses and their running as well as wider questions to do with the underlying philosophy that informed the practices of the spa and the treatments available. Interviews were accompanied by opportunities to look round the spas, meet other staff informally and undertake limited personal observations of facilities and treatment spaces. Finally, in addition to the questionnaire and interviews, some data were drawn from the analysis of textual material (advertising, promotion and magazine features) on spas.

Spaces of the Spa

As noted above, research on therapeutic landscapes has developed beyond a simple focus on the environments of healing spaces to incorporate recognition of therapeutic practices associated with such landscapes and of the interaction between people and place as part of the healing process. Clearly, however, space and place are still central to the experience of therapeutic landscapes and to the wellbeing practices and therapies that are enacted. The two spas that featured in this research were (so the popular literature and advertising would suggest) typical in drawing heavily on their location, not only as an important site for the situating the spa but also as central to the therapeutic benefits that could be offered. Both made reference to the 'beautiful environment' of the spa (both were located in attractive countryside settings with the Devon spa, in particular, surrounded by spectacular scenery) and also to the ways in which the rural environment enhanced the therapeutic practices through creating a sense of wellbeing. Frequent reference was made in the interviews and in the promotional literature of the spas to the 'relaxing' and 'nurturing' properties of the spas and their settings. Such references are particularly important in the light of the contemporary emphasis on relaxation as a crucial to (especially women's) health and also to the wider links between mental and physical wellbeing as discussed above. As one spa manager explained:

> The setting is fabulous, especially for the pamper breaks. They are invariably inland people coming to the coast. … We wish to provide an environment that is relaxing, re-energising. It's a nurturing environment and it allows you to escape. (Devon spa)

Similarly, the importance of the spa setting as part of the therapeutic treatments was emphasized in a review of the 'Top Ten Luxury Spas in the UK' in 'Homes and Gardens' magazine (2008).

Treatments at D.... Health Spa are superb and the stunning sea views are a tonic in themselves.

You don't know whether you're asleep, awake or floating along the North Devon coastline just outside. A guest cried as she admitted she'd never felt so nurtured.

The following quote also illustrates how the nurturing capacity of the spa environment contributed to its rejuvenating properties in terms the link between physical illness, stress and mental health.

I'd love to get the doctors round here to start to realize the value of what it is we can provide for people... we have, quite regularly, people who are grieving brought along by friends and it's a very nurturing environment. We have examples of women who go through holistic relaxation therapy and they will be in floods of tears through the physical process that is taking place – releasing tension – better than a sedative any day.... I think there are obvious social benefits that come out of the environment we are creating. (Devon spa)

In addition to the nurturing/relaxing affects of the environment, the inside space of the spa was also seen to contribute to its therapeutic properties. Managers of both spas emphasized the importance of providing an environment in which visitors felt comfortable. 'Hanging out' in the spa, beside the pool or in the lounge, was seen to be part of the restorative value of the spa and managers recognized the need to protect that space. It was also important that the space is preserved as an 'escape' in which spa users could distance themselves from the stresses of their daily lives. The questionnaire demonstrated the important role the spa filled in providing visitors with a place away from 'normal' responsibilities which they could have 'for themselves'. Again, this sentiment chimes both with earlier points about the value of relaxation and escape as part of the therapeutic process and with, as expanded on below, the role of pampering within contemporary (gendered) ideas of wellbeing.

There are other issues that could be explored in relation to the space of the spa. Links can be made, for example, with the notion of bodily discipline as discussed below. In both spas the managers spoke about the management of the different internal areas of the spa and of the regulation of bodies using those areas. Alternatively, the space of the spa was also remarked on in discussions of nature – again emphasizing the association between the location and environment of the spa and the practices and treatments that were supported. The following section, however, moves on to explore the spa in the context of the relationship between gender identity, wellbeing and leisure.

Spas, Leisure and Gender Identity

This chapter, as discussed in the context of recent literature on gender and wellbeing, positions the spa within an increasingly important relationship between contemporary gender identity (particularly femininity), the body, leisure and health. It argues, in so doing, that the kind of therapeutic space provided by the spa responds to (and is created by) new femininities in which women's wellbeing and ability to perform requires, and justifies, bodily and mental relaxation, indulgence and escape. It also asserts the importance of particular expectations of bodily size, shape and appearance to ideas of fitness and health. Thus the research sought to establish some basic understanding of the nature of spa use and of the expectations of spa visitors.

The spa as a regular, or at least unexceptional, part of contemporary leisure and beauty routines appeared to be born out by the research; on the evidence of the spas I visited, demand is high. The Gloucestershire spa has a membership capacity of 500 – it was currently full with a further 570 on a waiting list. The Devon spa has no fixed membership but takes visitors for days or short breaks – this spa was also reporting a rapid growth in demand.[1] There was also evidence of considerable unexploited demand; 78 per cent of those completing the questionnaire claimed that they would like to visit the spa more frequently if they could. In total 80 per cent of spa users said that their visits were of high or medium priority in the context of their daily lives. The use of the spa is very clearly and significantly gendered – 85 per cent of those responding to the spa questionnaire were women – a finding that is echoed in evidence collected elsewhere. A survey of spas in the US, for example, concluded that 71 per cent of all spa goers in 2003 were women (McNeil and Ragins 2005).

While spa visits are not exclusive, they are expensive and, for many, a luxury. But despite this the visitors were not all in high paying jobs (although over 70 per cent were in employment). Amongst the occupations listed by visitors were professional jobs such as IT management consultants, teachers, and GPs, but there were also non-professional jobs such as carers, hairdressers and cleaners listed. The spa managers in Devon were very conscious of the widening appeal of the spa and spoke of the very diverse range of visitors they received:

> We attract completely different people from what we were expecting. We are not for the mega rich – we have just ordinary people most of the time who want a treat.... You're getting people who are earning £6.00, £6.50 an hour and they'll be coming and spending £100, possibly £150 for a day.

The size and the varied nature of the client base demonstrates the wide appeal that spas now appear to hold, particularly for women. Examining the reasons behind this popularity informs the ways in which spas reflect and reinforce aspects of contemporary gender identities – particularly in relation to ideas of reward and retreat and to the contradictory ways in which the body is drawn into both the discipline and empowerment of women in conforming to the expectations

---

1   It should be noted that the research was undertaken prior to the recent economic downturn in the UK and this may be reflected in spa visits.

surrounding gender identities. When asked why they visited the spa, most questionnaire respondents gave several reasons including health and fitness, the beauty treatments and meeting friends. More important, however, were relaxation, escape and 'pampering'. Almost 80 per cent of respondents included 'relaxation' as a reason for visiting the spa and more cited 'pampering' than 'fitness'. For 21 per cent the *main* purpose of their spa visits was to be pampered. What was clear from the questionnaire responses and from the interviews was the sense in which the spa provided an opportunity for visitors to have time to themselves, to engage in legitimate relaxation in a space that took them away from their everyday lives and responsibilities. This was, according to one respondent, '*guilt-free relaxation*'.

'Pampering' appeared to add an extra dimension to the relaxation by adding a sense of luxury and indulgence. It also created the sense of the spa as some kind of hybrid, sitting somewhere between a routinized and familiar part of everyday life and a special treat. Pampering was clearly seen as a feminized concept, stressing the role of the spa as a regular and important part of women's gender identity. Earning a 'pamper day' at the spa can be related to birthdays or other life events but is also aimed at celebrating achievements and also coming through difficult times. Specifically, it provides a way of rewarding women's routine and everyday activities (especially around caring for the family) while reinforcing the notion that relaxation and escape, are for women, a treat.

The pamper day as a treat is evidenced in the fact that many spa guests have received their day as a gift in celebration or thanks. According to the questionnaire survey, 40 per cent of all women using the spa had been given a pamper day as a gift – in contrast none of the male respondents had received such a gift. Pampering stresses the notion of escape as essential to the maintenance of a 'balanced' life and suggests that luxury and indulgence help women to manage the mundane and stressful aspects of life.

> There is no question, it (the spa) does relax. From simple things like taking people out of their everyday lives and putting them in an environment where we don't have mobile phones – cups of tea and coffee are brought to you and somebody is running around looking after you. Just that alone is a lovely thing to have. (Devon spa)

As a gift or reward, however routinized and mundane, a spa pamper day gives women some time to themselves, away from their everyday roles. But although such days can allow women to assert their needs and independence, they are designed to maintain rather than challenge existing gender identities. As noted earlier, feminist writers have made similar arguments in relation to cosmetic surgery, often disagreeing as to whether body alteration through cosmetic surgery represents social control over women's bodies or a source of empowerment for women wishing to change their appearance (see Davis 2003, Petersen 2007). To some extent pampering highlights the same issues, namely the conflict between empowerment and control. Some suggest that pampering, while providing a sense of luxury, escape and wellbeing may also/rather be seen as a device to maintain established gender relations – its temporary and transitory nature being a tactic for sustaining the status quo and disciplining the body (see Black and Sharma 2001).

Spas and the Disciplined Body

Foucault's work on the 'care of the self' is, as noted above, useful in thinking through the ways in which the practices of the spa can be seen as part of a process of improvement in which the individual engages with 'the aesthetics of existence' in a series of treatments and techniques that are self-affirming and nurturing. In this final section of the chapter I suggest that such practices of wellbeing also serve to regulate the body. In so doing they reflect a sense in which wellbeing incorporates a sense of conformity to certain bodily expectations and the ability to regulate the self in accordance with physical norms. Elsewhere other authors have suggested that such an approach makes connections between the care of the self and Foucault's earlier work on self-discipline (see Parr 2002, Straughan 2010).

Thus while emphasis may be placed on relaxation and pampering of the body, there is still an underlying value system which assumes and rewards a particular kind of body. This is apparent in the continuing role of beauty treatments within the spa and highly traditional assumptions about the demands and priorities of clients. The women spa users in particular seemed to have an ambivalent response to the more energetic fitness activities although 27 respondents did claim to use the gym and fitness studio.

This was a lower number, however, than used the pool (46), hot tub (39) or 'beauty' treatments (36). Clearly, with many of the respondents claiming 'fitness' was a reason for visiting the spa, this was seen as a part of or complementary to rather than in conflict with, pampering. Fitness, then, for the majority of spa users questioned here seemed to be a holistic notion and prioritized relaxation as central to the healthy body.

For women in particular it seemed that the fit body is achieved through holistic therapy directed as much at the mind and emotions as the physical form of the body. Practices of pampering and relaxation were, however, not seen as ways of 'letting the body go' or indulging the body in a way that conflicted with stereo-typical, and highly gendered, notions of the acceptable body. So, in both spas, treatments and therapies aimed at pampering and relaxation were accompanied by nutritious and modest snacks and advice on weight loss and 'healthy eating'.

Highly important to the ideas of self-discipline within the spa was the emphasis placed on the beauty therapies as a central part of the therapeutic experience and of overall wellbeing. Such therapies were aimed almost exclusively at women in both spas and included some very conventional treatments as well as more holistic and 'natural' practices. Treatments were very well used and popular – only 10 per cent questionnaire respondents had *not* had one of the beauty treatments offered – and in both spas they were seen as a growing and indispensable part of the overall business, responsible for a sizable part of the business turnover in each case. Spa managers were very keen to respond to client's demands where beauty therapies were concerned. In the Gloucestershire spa it was explained that they had recently started up a 'Botox clinic' which was 'there in case somebody needed it'. The Devon spa claimed to have introduced certain treatments (manicure and leg waxing) for those clients ('typically career women getting away on holiday') who 'normally had them elsewhere but had not had time' before coming away.

While there was an assumption that clients were striving for a particular kind of body, both spas were keen to assert their distance from the conventional beauty salons. As one explained:

> We approach our beauty therapies from a holistic standpoint. Other spas are much more beauty orientated … a lot of painted bimbos who are doing beauty therapies. Someone who is coming to us for a facial is not having what they have in a beauty salon… we don't give, for example, a facial that's less than an hour. (Devon Spa)

Despite this claim about the way beauty treatments are undertaken in the spa, there is still an acknowledgment that such treatments are an accepted and routinized part of the maintenance of a healthy and fit body for women. Comments were made by respondents and spa managers about the link between feeling good and looking good and to the role of not only relaxation and nurturing but also cosmetic treatments in enhancing that link. It was also acknowledged that technology could be important in the process of incorporating beauty treatments in wellbeing practices. Again, while quick to defend the 'holistic' and 'natural' properties of spa therapies, spa managers in both spas clearly saw no contradiction in the use of (sometimes quite sophisticated) technology in the form of equipment for certain treatments or exercises and in cosmetic products. There is no space in this chapter to develop these ideas about the blurring of boundaries between nature and technology in holistic therapies but it may form an interesting direction for further work in the context of wellbeing and therapeutic geographies.

Finally, it is important to make brief reference here, in the context of bodily discipline to the ways in which the internal spaces of the spa are managed and regulated. In both spas the managers saw their market success as partly related to their sensitivity to the need to protect the quiet, relaxed atmosphere and to retain the nurturing and rejuvenating qualities provided by the calm environment.

> When my ladies are sitting round in towelling robes they don't want a bloke to walk in and start exercising… most hotels that run pamper days have this problem – people staying in the hotel can just drift in and it does change what you are doing. You need to separate the spa facilities from the drop-in facilities otherwise you change the ambience. (Gloucestershire spa)

The management of the space of the spa in this way can also be seen as a way of controlling and disciplining the body. There is no doubt that spa users, in both cases studied, were carefully controlled in order to protect the 'right' atmosphere. So as one manager reported:

> I have evicted one member who behaved inappropriately. I told a further two that if they step over the line one more time they'll be gone. I have no qualms about doing that.

Conclusion

This chapter has provided an original perspective on the notion of the therapeutic landscape in a discussion of the contemporary spa. It has argued that the spa plays an important role as a leisure space and needs to be examined as an increasingly regular and high profile element of health and leisure routines, particularly amongst women. Spa visits occupy an interesting position in clients' leisure routines as somewhere between a regular part of their leisure experience and a one-off treat.

Discussion of the spa as a therapeutic space has made reference to Foucault's ideas of the care of the self in the context of both nurturing and disciplinary practices of wellbeing. Such ideas position the relaxation and recovery therapies of the spa within contemporary leisure routines and embodied identities. They acknowledge the role of emotional wellbeing, of escape and pampering as central to the health of many women and to the individual's project of self-improvement and self affirmation. They also show these therapies to reinforce strong expectations of appropriate size, shape and appearance within notions of the fit body. The luxury and indulgence provided by the spa thus also serves to regulate the body and by stressing the relationship between fitness, wellbeing and particular body form helps to further justify women's spa visits.

The care of the self also provides an important way of linking the health and beauty aspects of the spa and thinking about the use of medical and scientific discourses/expertise in spa therapies. There are parallels here with the recent work of Straughan (2010) in which she examines the beauty salon as a clinic arguing that the present-day beauty salon emphasizes the scientific status of treatments and products. Straughan suggests that customers are subject to a 'medical gaze' in which the body is fragmented and becomes a site for concern in need of particular therapies. There are many elements of the spa which straddle this border between the medical and the cosmetic and which can be understood as part of the rather fluid and shifting relationship between wellbeing, fitness and self-regulation. While this chapter has begun to explore such issues as relevant to the spa as a therapeutic space, there are opportunities for further development in future research.

Finally, the chapter has also very briefly noted the importance of spa therapy as a 'natural' process of healing/wellbeing. It has drawn attention to the use of nature in both the spaces of the spa (in reference to more traditional perspectives on therapeutic landscapes) and in the practices and performances of spa therapies. The chapter has acknowledged the ways in which such spaces and therapies challenge the divisions between nature and technology. Clearly these challenges are also relevant to the body itself, to debates surrounding the relationship between the biological and the social body and to the ways in which cosmetic practices can be employed in shaping the idea and the reality of the 'healthy body'. Again, there is considerable scope here for the development of these ideas in relation to future research on both therapeutic geographies and the spa.

References

Bahktin, M. 1984. *Rebelais and His World*. Bloomington, IN: Indiana University Press.

Black, P. 2002. 'Ordinary people come through here': Locating the beauty salon in women's lives. *Feminist Review*, 71, 2-17.

Black, P. 2004. *The Body Industry: Gender, Culture, Pleasure*. London: Routledge.

Black, P. and Sharma, U. 2001. Men are real, Women are 'made up': Beauty therapy and the construction of femininity. *The Sociological Review*, 49(1), 100-16.

Cayleff, S. 1988. Gender, ideology and the water-cure movement, in *Other Healers: Unorthodox Medicine in America*, edited by N Gevitz. Baltimore, MD: John Hopkins University Press, 82-98.

Conradson, D. 2005. Landscape, care and the relational self: Therapeutic encounters in rural England. *Health and Place*, 11, 337-48.

Curtis, S., Gesler, W., Fabian, K., Francis, S. and Priebe, S. 2007. Therapeutic landscapes in hospital design: A qualitative assessment by staff and service users of the design of a new mental health inpatient unit. *Environment and Planning C: Government and Policy*, 25, 591-610.

Davis, K. 2003. *Dubious Equalities and Embodied Differences: Cultural Studies on Cosmetic Surgery*. Oxford: Rowman & Littlefield.

Foley, R. 2010. *Healing Waters: Therapeutic Landscapes in Historic and Contemporary Ireland*. Farnham: Ashgate.

Gesler, W. 1998. Bath's reputation as a healing place, in *Putting Health into Place: Landscape, Identity and Wellbeing*, edited by R. Kearnes and W. Gesler. Syracuse, NY: Syracuse University Press, 17-35.

Holliday, R. and Thompson, G. 2001. A body of work, in *Contested Bodies*, edited by R. Holliday and J. Hassard. London: Routledge, 117-33.

Homes and Gardens. 2008. Top Ten Luxury Spas in the UK. *Homes and Gardens*. January 2008.

Kearns, R. and Gesler, W. 1998. *Putting Health into Place: Landscape, Identity and Wellbeing*. Syracuse, NY: Syracuse University Press.

Lea, J. 2008. Retreating to nature: Rethinking 'therapeutic landscapes'. *Area*, 40(1), 90-98.

Lea, J. 2009. Liberation or limitation? Understanding Iyengar yoga as a practice of the self. *Body and Society*, 15, 71-92.

McNeil, K. and Ragins, E. 2005. Staying in the spa marketing game: Trends, challenges, strategies and techniques. *Journal of Vacation Marketing*, 11, 31-9.

Milligan, C. 2007. Restoration or Risk? Exploring the Place of the Common Place, in *Therapeutic Landscapes*, edited by A. Williams. Aldershot: Ashgate, 255-72.

Moore, S. 2010. Is the healthy body gendered? Towards a critique of the new paradigm of health. *Body and Society*, 16, 95-118.

Parr, H. 2002. New body-geographies: The embodied spaces of health and medical information on the Internet. *Environment and Planning D: Society and Space*, 20, 73-95.

Parr, H. 2007. Mental health, nature work and social inclusion. *Environment and Planning D: Society and Space*, 25, 537-61.

Petersen, A. 2007. *The Body in Question: A Socio-cultural Approach*. London: Routledge.

Philo, C. 2000. The birth of the clinic: An unknown work of medical geography. *Area*, 30, 11-19.

Pitts-Taylor, V. 2007. *Surgery Junkies: Wellness and Pathology in Cosmetic Culture*. New Jersey: Rutgers University Press.

Straughan, E. 2010. The salon as clinic: Problematising, treating and caring for skin. *Social and Cultural Geography*, 11(7), 647-59.

Williams, A. 2007. *Therapeutic Landscapes: Geographies of Health*. Aldershot: Ashgate.

Williams, A. 1999. *Therapeutic Landscapes: The Dynamic Between Place and Wellness*. Lanham, MD: University Press of America.

Chapter 15

Place, Place-making and Planning: An Integral Perspective with Wellbeing in (Body) Mind (and Spirit)

Ian Wight

Wellbeing and place have in common an essential wholeness, goodness and togetherness, a commonality resisting easy deconstruction, abhorring insensitive reduction and confounding simplistic quantification. They both transcend individual disciplines and professions; they are, above all, integrations of all those qualities most valued: the good, the true and the beautiful. They range wide and deep; they implicate our whole being – body, mind, soul and spirit. One of these, place-making, is perhaps more means than end, the other, wellbeing, more end than means, but they are indubitably linked. Although we apprehend this connection from everyday, and extraordinary, experience, a shift in perspective seems required to fully see and feel, and validate, this. Wellbeing and place have a primalcy and potency that demands we make this effort, in service to our selves, both individually and collectively, if we are not simply to survive but to thrive, if we are not doomed to languish but to flourish. Wellbeing and place cannot simply be studied as objective 'its'; they are an integration of integrations in flux as dynamic enactions. And as such, our concern must be with the enacting that is place-making and with the enacting of wellbeing (as whole-making) – ongoing productions both, that might just be amenable to some enlightened planning, policy and design. But how? This chapter proposes that, at the very least we will need a more integral approach.

An integral perspective has most recently been associated with the work of Ken Wilber (1994, 1996, 1997, 1998, 2000a, 2000b, 2006, 2007), a current exponent of one of several lineages of integrationist thinking (Molz 2010) that attempts to transcend and include many perspectives not normally conjoined. Driven by its own 'wholing' and 'placing' impulses, an integral perspective is rooted in a concern to integrate objective *and* subjective realms, in their individual *and* collective dimensions (Crittenden 1997, Esbjorn-Hargens 2009). The approach also embraces multiple lines of development, such as those associated with different strands of wellbeing, and attends to different levels of development, such as the distinction in wellbeing circles between lower and higher flourishing (Taylor 2007, Vernon 2008). Place serves well to ground such theorizing, as an integration venue *par excellence*. Yet viewing place itself through an integral lens also expands notions of place, including an enhanced understanding of its changing status through time within both analytical and policy elaborations. From its origins in pre-modernity, through marginalization in modernity, and a partial rehabilitation in post-modernity (Wight 2002, 2005, 2006, 2011), the underlying dynamics of place have become more apparent and less easy to ignore. Informed by an integral approach, wellbeing may be conceptualized as a product of place-making, engaging body, mind and spirit – the hand, head and heart of some formulations (e.g. McIntosh 2008). Place, in turn, may be conceptualized through the quadrants of Wilber's integral approach as the integration of physicality, functionality, community and spirituality. If planning and the related policy and design can be re-conceptualized as integral place-making, they may come to play a more significant role in the fundamental public policy concern of delivering and sustaining wellbeing.

This chapter seeks to tease out aspects of the inter-relationship of place, place-making and planning, with wellbeing in mind – but also with regard for body and spirit, for enacting that is embodying and inspiring as well as mindful. It represents an attempt to explore integral theory in action, in service of more fully harnessing the combined potential of wellbeing and place. Beginning with some 'auto-poiesis' that seeks to 'place' this author at this time, the chapter then demonstrates the potential value of an integral perspective for attempts to ground wellbeing in place. This inter-relationship between the personal and the conceptual is pursued with a view to evoking a sense of an emerging planning that is both more conducive to embracing wellbeing and consonant with a refreshed encounter with the policy and design contexts. The chapter then moves to attempt an integral embrace that can address wellbeing more directly, albeit in a preliminary and speculative manner. This paves the way for a specific consideration of place-making and wellbeing (as whole-making) in combination. The chapter closes with a consideration of how these two concepts might be better 'meshed' in practice, through an emerging social technology, 'meshworking', which has seeming application not only for planning, but also for the realms of policy and design.

Auto-Poiesis: A Personal Placing

An integral perspective honours the first-person perspective, as much as the second- and third-person perspectives. Our 'I' cannot be repressed, especially where wellbeing inquiry is concerned. Wellbeing has only recently appeared on the horizon of my own interests. Place, by contrast, has been foundational and fundamental for as long back as I can remember, both exercising and actualizing my intellect. Place for me is primal and potent, timeless and never more timely. My emerging sense of the place/wellbeing nexus is one that happens to privilege place, but in a way that I feel embodies, ensouls and inspirits wellbeing and as such I hypothesize wellbeing as a manifestation of implacement (Schneekloth and Shibley 2000). But the place I have in mind is much more than simply geographical or locational; it is integral in the sense of including while transcending the pre-modern, modern and post-modern notions, or 'senses', of place. It also has an inherent dynamic in that place is constantly emerging, both developing and evolving, constituting a verb as much as a noun, in that place is always in the process of being made and remade in pursuit of an ever-greater sense of the good (and the true, and the beautiful). Could such integral place-making be synonymous with wellbeing as a socio-spiritual construction? Can they inspire one another, and generate a synergy for planning, policy and design to harness?

More pragmatically, I approach these questions as an applied geographer and a professional planner. While some of my early geography interests were more spatial, my focus is now very much 'placial'. I have practiced planning as the allocation, regulation and re-arrangement of space, but I currently advocate planning as place-making. Over the past decade or so, I have shifted increasingly towards an integral framing of my professional and academic endeavours. In exploring an integral approach to wellbeing and place, I build on earlier work that explored the application of integral theory to planning and place-making, an applied integral ecology (Wight 2002, 2005, 2006, 2011).

It is particularly heart-warming, and perhaps intriguing, to observe that in recent years the targets of policy have expanded beyond the purely material and economic to embrace more subjective dimensions of human flourishing. This indicates an anticipation of multiple perspectives needing to be incorporated across individual and collective scales of analysis which in turn indicates a possible openness to an integral perspective. Moreover, an integral perspective is particularly appropriate to examine the complex ways in which place and wellbeing interact. This interaction, to date, has remained relatively under-researched and under-theorized.

It should probably be stressed that the 'place' in view here (and possibly also the 'wellbeing' in view) is probably a good deal more capacious than that envisaged by many authors writing about wellbeing. Spatial containers and analytical scales can be too limiting, where integrally-conceived notions of place and wellbeing are being invoked. For example, an integral perspective naturally and necessarily includes a regard for spiritual considerations as part of 'the subjective dimensions of human flourishing'. Spirituality is greatly neglected in explorations of wellbeing, lurking as a kind of 'elephant in the room', a huge and significantly missing dimension. 'Place' then is consciously and deliberately conceptualized as the integration of physicality, functionality, community and spirituality and given an enacted/enacting form as place-making. Wellbeing is also interpreted, following Vernon (2008), as much more than happiness or even a meaning-filled life. Instead, consciously and deliberately, it is regarded as including the transcendent, the mysterious and, ultimately, a loving spirituality; it is given a dynamic, spiralling form, as 'whole-making'. Together, in combination, they involve a full engagement of body, mind and soul, potentially manifesting a palpable *poiesis.*

Grounding Wellbeing in Place: The Integral Perspective in Action

Although there are several lineages of integral thinking (Molz 2010), the integral approach employed here is rooted in the work of Ken Wilber (1994, 1996, 1997, 1998, 2000a, 2000b, 2006, 2007, Wilber et al. 2008) and particularly his 'integral' phase, sometimes referred to as 'Wilber-4'. At its simplest, an integral approach is 'comprehensive, balanced, and inclusive ... not leaving anything out' (Lundy 2010). The approach is structured through three intersecting dimensions which enable, and insist upon, attention to a multiplicitly of considerations. These dimensions comprise: four primary perspectives, termed the four *quadrants* of knowledge; several *levels* of development which connote increasing complexity; multiple *lines* of development, each of which involves stages of evolution or emergence. There are more complex variants of the approach which also include additional considerations such as *state* (as in states of consciousness) and *type* (such as personality or gender), but these are not examined in this introductory encounter.

Applications of the integral approach within contexts relevant to wellbeing and/or place have begun to appear recently including within the contexts of applied integral ecology (Wight 2005), consciousness and healing (Schlitz et al. 2005), human flourishing (Dacher 2006), social identity development (Quinones-Rosado 2007), whole systems health care (Schlitz 2008) and public health promotion (Hanlon and Carlisle 2010; Hanlon et al. 2010a, 2010b; Lundy 2010). Introducing the integral framework to new audiences remains a challenge; the approach entails large claims that warrant a large contextualization (McIntosh 2007), but here, as elsewhere, with limited space only the lightest of tastes can be offered in hope that appetites may be whetted for further independent exploration. The following introduction borrows heavily from one of the most recent applications (Lundy 2010) who notes that:

> The growing global success of the Integral approach lies in its capacity to address the full complexity of human experience in an increasingly complex world... The Integral map makes room for all forms of action and inquiry, and the evidence they generate... a map of reality that incorporates both subjective and objective dimensions of life, in individuals as well as in collective contexts. And it is a map of reality that accounts for human development throughout the lifecourse. (Lundy 2010: 46)

As indicated above, the main elements of the Integral map, for the purposes of this chapter, are those also highlighted by Lundy – quadrants, levels and lines:

> *QUADRANTS call our attention to four unique dimensions of any experience...* The Integral model takes into account subjective experience as well as objective experience, and pays equal attention to the individual and the group or community... demonstrates how these four dimensions are interconnected and irreducible. (Lundy 2010: 46)

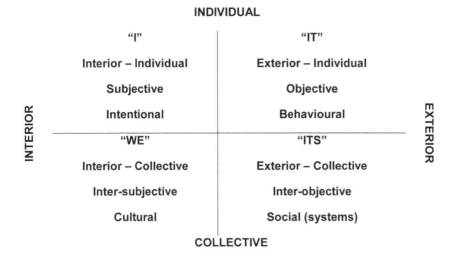

INDIVIDUAL

"I"	"IT"
Interior – Individual	Exterior – Individual
Subjective	Objective
Intentional	Behavioural
"WE"	"ITS"
Interior – Collective	Exterior – Collective
Inter-subjective	Inter-objective
Cultural	Social (systems)

INTERIOR ... EXTERIOR

COLLECTIVE

Figure 15.1   The Four Quadrants of Integral Theory

The four quadrants of the integral context are visualized in a simple 2 by 2 matrix (Figure 15.1), created by the intersection of an interior/exterior horizontal axis and an individual/collective vertical axis.

Lundy characterizes the levels of the integral approach as follows:

> *LEVELS of development call our attention to patterns of human growth and change*: The Integral model pays attention to human development, acknowledging patterns by which development unfolds throughout the lifecourse, from birth through old age. The patterns are mapped vertically as stages of development, or 'levels'. An important Integral insight is that healthy development is a core attribute of healthy people and healthy communities, and that each in turn influences the health of the natural environment. (Lundy 2010: 46-7)

One of the major examples of 'levels' in some integral analyses derives from Spiral Dynamics, an application of the developmental psychology of Clare Graves, popularized by the work of Beck and Cowan (1996). This work focuses on basic macro-values, referred to as V-memes that are manifest in both individuals and collectivities such as cultures and other large-scale systems. The V-memes constitute a 'never-ending spiral' (Roemischer 2002) of levels. The complex values involved in the set of V-memes are colour-coded for simplicity of communication. The set begins with beige (associated with survival) which is successively encompassed by more complex V-memes, each of which effectively transcend and include earlier V-Memes (see Figure 15.2). The first six levels are conceptualized as constituting a 'first-tier' macro-level of existence, generally associated with surviving or simply subsisting. A 'second-tier' of existence has

Level	Colour	Popular Name	Thinking	Organizing Code	Driving Principle
First Tier vMEMES (Subsisting)					
1.	Beige	Survival Sense	Instinctive	Survival Band	Instinct Driven
2.	Purple	KinSpirits	Animistic	Tribal Order	Safety Driven
3.	Red	PowerGods	Egocentric	Exploitive Empire	Power Driven
4.	Blue	TruthForce	Authority	Authority Structure	Order Driven
5.	Orange	StriveDrive	Strategic	Strategic Enterprise	Success Driven
6.	Green	HumanBond	Consensus	Social Network	People Driven
Second Tier vMEMES (Being)					
7.	Yellow	FlexFlow	Ecological	Systemic Flow	Process Oriented
8.	Turquoise	WholeView	Holistic	Holistic Organism	Synthesis

Figure 15.2 Spiral Dynamics V-memes Overview

Source: Based on workshop and conference presentation materials produced by Beck and Cowan, 1996

been identified as succeeding the first-tier, with an anticipated rolling out of a new series of V-memes, associated more with 'being', with comparative thriving and flourishing.

The integral approach is regarded as largely 'second-tier', engaging yellow and turquoise V-meme levels in the terminology of Spiral Dynamics. However, the approach is explicit in also specifically including the 'first-tier' V-memes; following the theory of spiral dynamics, the relationship is one of 'transcending while including'. Earlier V-memes may be associated with less complexity, but are also deemed more fundamental than later-emerging V-memes. However, each first-tier V-meme is exclusive in orientation, disparaging or denying earlier V-memes; in relation to these values there is an in-built resistance to an integral approach. The 'extra-ordinariness' of the second-tier set is associated with an extraordinary change and challenge in life conditions; the 'leap' from first- to second-tier is considered equivalent to all the change represented by the span of all six first-tier V-memes. It represents a massive change in world-view and, literally, in world-consciousness.

Some of the nature of this massive change is conveyed by another, simpler, three-level version of the Integral model, focusing on the stage-by-stage progression of human development from *self-centric* (often known as ego-centric) through *socio-centric* (or ethno-centric) to *world-centric*. As Lundy elaborates:

> Each marks a developmental level in which our capacity for care and concern becomes increasingly inclusive, beginning with 'me' at the self-centric stage, expanding to include 'us' (i.e. people like me) at the socio-centric stage and, finally, 'all of us' at the world-centric stage. With each evolving stage we expand our capacity to take deeper and wider perspectives into account. (Lundy 2010: 49)

An integral perspective is particularly associated with a world-centric, or greater, perspective-taking capacity. It also engages other more evolved or emerging perspectives, occasionally expressed, for example, as post-conventional, trans-personal, or post-formal. And it is associated with a level of cognitive development identified as 'vision-logic' thinking. These distinctions relate mostly to developing stages within particular 'lines' of development, a notion comparable to the 'multiple intelligences' proposed in the work of Howard Gardner (2006); such lines of development constitute the third key element of the overall model:

> *LINES of development call our attention to the multiple areas in which change and growth occur and can be actively supported*: The Integral model acknowledges that human development unfolds in relatively independent areas or 'lines' – cognitive, values, morals, needs, self-identity, emotions, interpersonal, etc. An individual may be highly developed in some areas, and less so in others. Becoming 'developmentally attentive' as individuals, as organizations, and as communities, we actively support healthy development in each of these areas. (Lundy 2010: 47)

This chapter then advocates a more explicit consideration of wellbeing and place in integral terms, an attempt to incorporate a regard that ranges across all the quadrants, all relevant levels and lines so as to better capture the complexities of the concepts of wellbeing and place. The endeavour in this direction in relation to place, placemaking and planning takes off rather directly from some previous work (Wight 2002, 2005, 2006). Wellbeing is given a more tentative and speculative treatment in a later section. In terms of Wilber's integral scheme, 'place' may be conceived most comprehensively and inclusively as ideally *all-quadrant. Place* spans exterior and interior, collective and individual, in exquisite balance, manifesting extra-ordinary 'co-related-ness'. Place merits consideration as a key venue for the integration of 'It/Its' and 'We' with the 'I' of the beholder in both material form and non-material consciousness. As well as multi-quadrant, place must also be conceived as *multi-level*. It has both primalcy and potency in that it integrates past, present and future. It is submergent and emergent which resonates with a common thread in Wilber's theorizations of 'nestwork' within his mobilization of the concept of the Great Nest of Being (Wilber 1996, 1998). Place is always in flux, in development, being made and re-made, to better ground and situate the development of individuals and collectivities. Its basic development structure is nest-like or spiral-form, including, whilst also transcending, less-developed, less complex stages, yet always anticipating more developed, more complex stages. Place has an evolutionary never-ending-ness (Roemischer 2002).

Within Wilber's scheme, place can be positioned and situated within emergent thinking on cognitive development around the transition from late formal cognition to early vision-logic. Place is on the verge of going 'post-formal' in the sense of a manifestation of a consciousness that is increasingly aware of itself, an emergent becoming, with an attention to its intention. This 'post-formality', or going beyond a past privileging of the exterior, the concrete and the physical attributes of place, provides the main contemporary context within which an emerging form of planning as integral place-making can logically be envisioned. Wilber himself observes a range of post-formal (vision-logic) cognition:

> These post-formal stages generally move beyond the formal/mechanistic phases (of early 'formop') into various stages of relativity, pluralistic systems, and contextualism (early vision-logic), and from there into stages of meta-systematic, integrated, unified, dialectical, and holistic thinking (middle to late vision-logic). This gives us a picture of the *highest mental domains* as being dynamic, developmental, dialectical, integrated. (Wilber 2000a: 22)

Conceptions of wellbeing as expressing higher flourishing (Taylor 2007, Vernon 2008) may be envisaged as reflecting a similar 'staging'. Higher flourishing has the feel of second-tier V-memes; lower flourishing, including manifestations such as happiness science (Ricard 2003, Layard 2007) or positive psychology (Seligman 2003), more obviously resonate with the realm of first-tier V-memes. Moreover, the 'higher' transcends and includes the 'lower'.

A final element in Wilber's theorization is his categorization of the world into four 'spheres': the physiosphere of the non-biological environment; the biosphere of organic life; the noosphere of sentient reflexivity; and the theosphere of spiritual encounter (Wilber, 1996). The noosphere, in transcending while including the physio-sphere and the bio-sphere, is likely to become engaged in a hitherto-unprecedented manner as a critical operative context for planning; increasingly, this will become *the* sphere of planning as placemaking and as conducing wellbeing as whole-making. Feelings, art and spirituality, defining qualities for the all-quadrant/all-level notion of place developed from Wilber's integral perspective, will similarly rise in planning consciousness. These qualities are not prominent in contemporary planning but the extent to which they remain foreign to planning may be indicative of how planning should expect to be regarded as foreign to most individuals. A greater emphasis on a place-making perspective may help to bring planning closer to a 'home' basis, or, at the very least, reduce the present disconnect.

Spiral Dynamics, discussed earlier, also offers a potential and significant bridge between Wilber's thought and contemporary planning in so far as it helps to place different 'plannings' in a developmental perspective. For example, a planning process may travel from *red* command-and control, to *blue* master planning and zoning, to *orange* strategic planning, to *green* communicative action and even possibly to *yellow/turquoise* ecological/wholistic 'second-tier' place-making and whole-making. To the extent that Spiral Dynamics can achieve an integral wholism, it may help usher in a planning that is in greater synchrony with place-making, particularly as the second-tier V-memes take deeper root in collective consciousness. Wilber made much of Spiral Dynamics in his past work, in part because it helped to advance his case for the constructive postmodernism that then underpinned his integral approach:

In the terms of Spiral Dynamics, the great strength of postmodernism is that it moved from orange scientific materialism to green pluralism, in a noble attempt to be more inclusive and sensitive to the marginalized others of rationality. But the downside of green pluralism is its subjectivity and relativism, which leaves the world splintered and fragmented... . And however important these multiple contexts are for moving beyond scientific materialism, if they become an end in themselves, they simply prevent the emergence of second-tier constructions, which will actually reweave the fragments in a global-holistic embrace. It is the emergence of this second-tier thinking upon which any truly integral model will depend – and this is the path of constructive postmodernism. (Wilber 2000b: 172)

In this, again, we glimpse the possible context for a more integral, constructively postmodern, second-tier planning with place-making in mind, literally and figuratively, individually and collectively. Perhaps most fundamentally, Wilber's perspective helps situate any better inter-relating of place, place-making and planning as 'a post-modernity project', or perhaps more accurately, a *post*-post-modernity project. But to what extent can planning be considered to be developing and evolving, to be occupying new worlds such as the post-post-modern and the post-formal? Planning may abstractly be conceived as a linking of thinking and acting (Friedmann 1987) in that it is both forward-thinking and intervention-oriented. It can take many forms, depending on the time/space setting and governing life-condition challenge to be addressed. It is possible to imagine planning as always evolving in terms of the level and stage of consciousness manifesting. For example, we may now be advancing on a period of post-conventional 'vision-logic' planning as distinct from a planning rooted in more conventional concrete/physical (conop-) or more formal instrumentally-rational (formop-) thinking.

Planning, at its best, has always been driven by an impulse to improve, to better, to do (more) good, to serve wellbeing in all contexts. At its core it is a linking endeavour that is linking appropriate knowledge and understanding with desired actions or good intentions. Planning as we know it, and especially in its current professional form, was born in modernity; mature professions and rational planning are possibly among the strongest markers of modernity. Planning has certainly aided social and cultural change and development, but within its own biases and predilections. In particular, planning has shown a space fixation, it has privileged the application of the sciences over the application of the arts and the humanities, it has focused on the physical, especially land and its use, or the environmental, and there has been a strong statutory bias which effectively privileges the *status quo*. Moreover, planning's essential strength for linking and integrating often seems to have become lost in the pragmatics and imperatives of 'action' that is 'practical' – as in habitual, status-quo-serving. Modern(ity) planning has been effectively a linking of 'first-tier' thought with a differentiating agenda and a segregating impulse, dominated by a concern with the 'IT' and 'ITS' worlds; it has been *anything but* integral.

Planning, along with policy and design, needs more consciously to become a part of a post-post-modernity project which can link 'second tier' thought with an integrating agenda and a meshing of all spheres, the so-called 'kosmopolitan' impulse, rooted in – until it is inevitably superceded – the method of an all-quadrant/all-level integral approach.. Such planning might be well-served by a renewed consideration of the central place of place, but place mobilized as an explicitly post-conventional (not pre-conventional) notion, rooted in a world-centric vision-logic rather than a nostalgic revival of ego-centric mythic/magical images and symbols. Integral planning, operationalized as post-conventional place-making, is thus 'logically-envisioned' as a necessary part of the post-post-modernity project that is increasingly promoted by integral thinkers.

The underlying evolutionary interest of integral thinking can be interpreted as going right to the heart of the process in and of planning. Intervention is interpreted as reflecting a wholeness-seeking imperative, involving the building of systems within systems within systems, but within the context of an integral embrace rather than a systems approach that privileges exterior elements. It also opens up a key sense of the evolving place of place, a conscience that is increasingly becoming consciously aware of itself; the context for a shift from the merely cosmopolitan to the explicitly 'kosmopolitan'. After making a strong, and essentially anchoring, appearance in pre-modernity, place seems to have become a casualty of *both* the success of modernity in achieving the differentiation of what Wilber describes as the Big Three, that is Self, Nature and Culture, *and* of modernity's failure in achieving their integration. But although modernity consigned place to a comparatively invisible role in (modern) planning (Casey 1997), it may now be enjoying something of a revival as part of the constructive post-post-modernity integration project that an integral approach promises. A progressively post-conventional notion of place may anchor a constructive postmodern rehabilitation and presage its continuation through the whole evolutionary spiral. Place can be considered to incorporate qualities essential to meeting basic and not-so-basic human needs; it is in a sense inseparable from being human. Yet place has too often too easily been dehumanized as simply an 'it', as an objective, often spatial, identifier, in what Wilber labels 'Flatland'. In terms of planning, the usage of 'place' in such a reductionist context is necessarily limited, and limiting; at best it marks a progressive modern planning, but one that only goes so far. This limitation bears testimony to the value of directly exploring a more expansive perspective on place. A fruitful manifestation of the integral approach conceives of place as *anything but a mere 'it'* and centrally includes consideration of the actual *making* of place in all its dynamics as the integration of 'It' and 'We' with the 'I' of the beholder.

Placing Wellbeing in an Integral Embrace

Wellbeing may be conceptualized as, at the very least, a by-product, if not a direct product, of the kind of integrally-informed place-making articulated in the previous section. If planning can be enacted as integral place-making, it cannot but play a more significant role in engendering greater wellbeing. But what if wellbeing itself is considered more directly through an integral lens? An integral wellbeing perspective may prove as generative as the integral place-making perspective bringing large dividends to the process of planning, policy and design.

In Wilber's framing of an integral approach, the quadrants elicit an appreciation that official and scientific discourse treats wellbeing simply and predominantly as an 'it', as something 'out there' rather than 'in here'. This dominant discourse privileges an objective exterior perspective to the detriment of one that is subjective and inter-subjective. In examining Wilber's levels, some important differentiations in the mobilizations of wellbeing begin to emerge, for example, between settling for a focus on happiness and striving to embrace something more profound such as meaning, mystery or love. Moreover, the distinction between lower and higher flourishing begins to make sense as a possible reflection of 'first-tier' happiness-like concerns versus 'second-tier' wellbeing, broadly and deeply conceived. The commendable efforts to tease out different dimensions of wellbeing such as the psychological, emotional, physical, spiritual and so forth, can be understood as an effort to distinguish different multiple 'lines' of development associated with wellbeing.

Reaching for a more 'wholistic' view, wellbeing evokes a highly generative interpretation through an integral lens. Conventional etymology needs to be bypassed; the definition of wellbeing as 'the state of being comfortable, healthy or happy' that is encountered in many dictionaries falls far short of the meanings possible through an integral approach. The more appropriate root word from an integral perspective is 'whole' which can be seen to encompass 'well' and much more besides. This reframing more easily invokes whole-system conceptualizations, including such schemes as the Medicine Wheel (Quinones-Rosado 2007). It also suggests a living, life-affirming orientation in favour of 'wholing', of seeking ever-greater, ever-more-exquisite, wholeness in ourselves, in our relationships and in our environments. We can comprehend 'wholing' as a deeper and wider form of 'healing': it is the action verb to the wellbeing noun in which enacting wellbeing becomes a form of 'wholing', of making more whole, of 'whole-making'. Enacting wellbeing seeks always to conduce, extend and embed wholeness as wellness, or thriving as flourishing; it is the 'wholing' practice that yields wellbeing.

This notion of wellbeing as 'wholing' or 'whole-making' opens up possibilities well beyond end-points such as happiness. Moreover, this interpretation engages our need for deep meaning beyond everyday life. It also encompasses the sense of being part of something bigger, more transcendent, mysterious, but loving which constitutes a spiritual dimension to wellbeing (Vernon 2008). And within an integral perspective there is room not only for the rational and the mystical, but also for the magical and mythical. With a focus on *wellbeing*, not ill-being, this framing as 'wholing' recognizes a developmental and evolutionary dynamic in a continuous ever-seeking to become ever-more whole.

Combining Place-making and Wellbeing (as Whole-Making)

In this chapter place-*making* (Schneekloth and Shibley 1994, 2000, Wight 2002, 2005, 2006) has been privileged over a focus on place per se. It is the making, and re-making, by the people in and of a place that is central, whatever the context or scale. The *making* of place, rather than place per se, is very deliberately stressed here to emphasize the application to practice. There is a considerable literature on 'place' but much less on 'place-making'. Early formulations may be traced to Australia in the work of Dovey (1985) or Winikoff (1995) and reflected in North America by Schneekloth and Shibley (1994, 2000). Place-making is mostly about collective action on common concerns in a concerted fashion. Specifically, the emphasis is on the making and the remaking as a process of taking action together and building momentum that can be both always complete and also never finished (Quayle and Driessen van der Lieck 1997). Place-making involves a positive trajectory towards the good, the true and the beautiful, healing in the process in the sense of 'wholing' and enacting wellbeing. However, the root notion of 'place' is also considered to represent much more than geography, much more than space, or space-time. Place simultaneously engages the physical, the functional, the communal and the spiritual and is a dynamically evolving notion. And as already discussed, an integral sense of place, and its making, transcends, while including, the best aspects of pre-modern, modern and post-modern 'senses' of place.

As we have seen in the previous section, wellbeing also gains particular and special interpretation when viewed through an integral lens, but this raises the question of how such a special formulation can be actualized, especially in the inter-relationships of place-making and wellbeing. Combining place-making and wellbeing in integral terms invokes meshworking, a very new notion which draws partly on the early integral wave but which also has roots in brain science (Hamilton 2008, 2010). The concept builds on the more familiar ground of networking, but develops this into an almost unrecognizable formulation. The attraction of this concept of meshworking is that it can respond to the challenge central to an integral approach of achieving uncommon 'integratedness' in real-time. Meshworking provides the context for operationalizing trans-disciplinary, trans-professional, trans-sectoral endeavour. Moreover, meshworking is described as both inherently ecological, and highly participatory (Torbert and Reason 2001), solidly biased in favour of conducing collective action, following the raising of the collective consciousness of all the players in an issue. Meshworking can be seen as ecological interconnectedness 'personified', in a complex collective collaborative context. To the extent that all of this can be contained, an associated integral container would be shaped through a primarily horizontal place-making which is rooted in the quadrants and valuing the balancing of interior and exterior, and a primarily vertical wellbeing which is rooted in the levels and lines and valuing the interaction of individual agency and collective communion. Wellbeing would seem to have its origins in the perennial philosophy, what Wilber termed the Great Chain of Being (Wilber 1996, 1998) while place-making reframes and reconstitutes such a chain more as a 'Nest'. Meshworking is a catalysing agent on a grand scale for an integral interaction and an integrated enaction.

Meshworking Placemaking and Wellbeing

The underlying planning, policy and design challenge may be expressed in terms of meshing place-making and wellbeing, as whole-making. An explicitly integrally-informed 'meshing' may be characterized as meshworking, a form of second-tier integral collaboration which comprises far more than mere cooperation or first-tier networking. This approach can take many scientists, professionals and citizens well outside their comfort zone, in part because it also necessitates them consciously going well *inside* themselves. An integral engagement of wellbeing and place-making entails engaging the ineffable, within the realms of consciousness, as much as it entails engaging exterior concrete form. It is an inner work project, engaging not just the mind, but body and spirit. This may be characterized as 'enaction inquiry', a form of advanced action research, blending first-, second- and third-person inquiry to unearth relevant subjective and inter-subjective knowledge as a complement to the otherwise dominance of objective knowledge (Torbert et al. 2004). This approach was piloted in a conference presentation which began with some first-person self-inquiry. Participants were invited to investigate two lines of questioning. Thinking of their experiences over the recent past, the questions posed were:

1. When do you know you are well, are – in fact – a well being?
2. When do you know you are in your place – your prime, thick place?

After some individual reflection on their responses, dyads were formed to facilitate a simple second-person co-inquiry, seeking points in common about wellbeing and place, and/or their relationship. A final plenary sharing aimed at some third-person synthesis of points which had potentially wider general significance, including the challenge, together, to consider any possible 'elephants in the room' that is something that we, some or all, might have been overlooking or avoiding in our reflections. Conferring was enabled; some communing was achieved; a common place was convened, however short-lived; wellbeing was enhanced and we left a little more 'whole', both individually and collectively. Whilst this represented a small experimental moment, this kind of process of inquiry that is collective, consultative, collaborative, cooperative can enable the scope and depth of reflection that is essential to activate an integral approach to the meshing of wellbeing and place-making.

My own efforts to enact the sentiments associated with an integral perspective have mostly focused on efforts to re-frame planning as place-making as a critique of the traditional space-fixation and which can move well beyond a concern with simply land and its use in order to effectively implicate planning in the much larger project of conducing wellbeing, as whole-making. The efforts have included programming a national planning conference (CIP 2008) on the theme of 'Planning by Design in Community – Making Great Places'. This entailed an effort to better integrate planning and design, through place-making in community, by problematizing a space focus in favour of a shift to a place focus. The conference explored how planning as place-making might involve people and planners working together to co-design space-place transformation, reflecting an integrally-informed view of places, especially in contrast with spaces (Wight 2011). In 2008, during academic leave in the UK, I began to notice the prominence of wellbeing in much public and professional discourse. I sought to explore coupling my long-established interest in place-making with my new interest in wellbeing. The work presented here marks a first effort to connect the two from an integral perspective. At first I was challenged to understand the essence of wellbeing. It had a different feel from my sense of place-making. My early journaling indicates a time when I was forced to conclude, in comparison with place-making, that: 'wellbeing ... just *is* ... it's subjective, it's inter-subjective... NOT made, NOT constructed'. Then I encountered Mark Vernon's *Wellbeing* (2008) and began to sense some integral connections, especially at the level of 'higher flourishing... the larger perspective in life':

> It is prior to lower flourishing because it informs and shapes the humdrum. It provides a sense of intrinsic meaning or overall direction or deeper purpose. It originates not in daily activities but in ethics, spirituality or religion... It is not just a concern with the piecemeal constituents of a good life, but a love of the good itself and a search for that good in life... It is characterized as a transcendent commitment, that if not imagined as belonging to another world is felt as a pull towards something that is deeper than or beyond the concerns that an individual would otherwise have for himself or herself... higher flourishing has the magnetic allure of what is good in itself. It is about the spirit level. (Vernon 2008: 6-7)

Vernon's insights on the essence of wellbeing sought to put a spiritual dimension into our understandings, as a foregrounded and central element:

> ... if our wellbeing depends in some way on that which is beyond us – or at the very least draws us to a state of purpose or serenity as if from the outside – it is by definition in large part unfamiliar, unusual, or unknown. It emerges as something shown or revealed, not told or made. It is an experience not a rule; although informed by rationality it outstrips rationality. (Vernon 2008: 12)

The 'well' in wellbeing goes back to the original notion of 'whole', when whole very much referenced body, mind, soul and spirit. This sense of wholeness constitutes some dignified pre-modernity that deserves to be preserved and sensitively integrated into contemporary thinking and practice. The coupling with place-making, and an integral perspective helps to render wellbeing as a form of 'whole-making' and it is in this combination that we might all find our post-postmodern calling, our co-mission-ing. Paraphrasing Martin Luther King, I have a dream, of professions, and an academy, of servant-leaders, as a community of wellbeings, striving above all for the wellbeing of all, in well-loved places: whole beings, in whole places, tending not just to inanimate matter, but to all that matters, in body, mind, soul and spirit.

References

Beck, D.E. and Cowan, C.C. 1996. *Spiral Dynamics: Mastering Values, Leadership, and Change*. Malden, MA: Blackwell Publishing.

Canadian Institute of Planners (CIP) 2008. *Planning by Design in Community: Making Great Places?* Canadian Institute of Planners Annual National Conference, Winnipeg, July 2008.

Casey, E.S. 1997. *The Fate of Place: A Philosophical History*. London: University of California Press.

Crittenden, J. 1997. What is the Meaning of 'Integral'?, in *The Eye of Spirit: An Integral Vision for a World Gone Slightly Mad*. Boston, MA: Shambhala, vii-xii.

Dacher, E.S. 2006. *Integral Health: The Path to Human Flourishing*. Laguna Beach, CA: Basic Health Publications Inc.

Dovey, K. 1985. An Ecology of Place and Placemaking: Structures, processes, knots of meaning, in *Place and Placemaking: Proceedings of the PAPER 85 Conference*, edited by K. Dovey, P. Downton and G. Missingham. Melbourne: Association of People and Physical Environment Research, 93-109.

Esbjorn-Hargens, S. 2009. *An Overview of Integral Theory: An All-inclusive Framework for the 21st Century* [Online]. Available from: www.dialogue4health. org/pdfs/3_18_09/E_H_Overview-IT.pdf [accessed: 25 May 2011].

Friedmann, J. 1987. *Planning in the Public Domain: From Knowledge to Action*. Princeton, NJ: Princeton University Press.

Gardner, H. 2006. *Multiple Intelligences: New Horizons in Theory and Practice*. 2nd Edition. New York: Basic Books.

Hamilton, M. 2008. *Integral City: Evolutionary Intelligences for the Human Hive*. Gabriola Island: New Society Publishers.

Hamilton, M. 2010. Meshworking Integral Intelligences for Resilient Environments: Enabling Order and Creativity in the Human Hive. Paper to the Integral Theory Conference: JFK University, Pleasant Hill CA, August 2010.

Hanlon, P. and Carlisle S. 2010. *Applying the Integral Model to Public Health* [Online]. Available at: www.afternow.org.uk/papers [accessed: 25 May 2011].

Hanlon, P., Carlisle, S., Reilly D., Lyon A. and Hannah, M. 2010a. *Applying the Integral Framework to the Problem of Obesity* [Online]. Available at: www. afternow.org.uk/papers [accessed: 25 May 2011].

Hanlon, P., Carlisle S., Reilly D., Lyon A. and Hannah, M. 2010b. Enabling Wellbeing in a Time of Radical Change: Integrative Public Health for the 21st Century. *Public Health*, 124(6), 305-12.

Layard, R. 2007. *Happiness: Lessons for a New Science*. Harmondsworth: Penguin.

Lundy, T. 2010. A paradigm to guide health promotion into the 21st century: The integral idea whose time has come. *Global Health Promotion*, 17(3), 44-53.

McIntosh, A. 2008. *Rekindling Community: Connecting People, Environment and Spirituality*. Totnes: Green Books Ltd.

McIntosh, S. 2007. *Integral Consciousness and the Future of Evolution: How the Integral Worldview is Transforming Politics, Culture and Spirituality*. St. Paul: Paragon House.

Molz, M. 2010. *The Many Faces of Integral: Towards a Reflexive Genealogy of Streams and a Dialogical Ecology of Voices*. Paper to the Integral Theory Conference: JFK University, Pleasant Hill CA, August 1, 2010.

Quayle, M. and Driessen van der Lieck, T. 1997. Growing Community: A Case for Hybrid Landscapes. *Landscape and Urban Planning*, 39, 99-107.

Quinones-Rosado, R. 2007. *Consciousness-in-Action: Toward an Integral Psychology of Liberation and Transformation*. Caguas, Puerto Rico: Ile Publications.

Ricard, M. 2003. *Happiness: A Guide to Developing Life's Most Important Skills*. London: Atlantic.

Roemischer, J. 2002. The Never-ending Upward Quest: The Practical and Spiritual Wisdom of Spiral Dynamics (An Interview with Don Beck). *What is Enlightenment?* Issue 22, Fall/Winter 2002, 105-26.

Schlitz, M.M. 2008. The Integral Model: Answering the Call for Whole Systems Health Care. *The Permanente Journal*, 12(2), 61-8.

Schlitz, M.M., Amorok, T. and Micozzi, M. 2005. *Consciousness and Healing: Integral Approaches to Mind-Body Medicine*. St. Louis, MO: Churchill Livingstone.

Schneekloth, L. and Shibley, R. 1994. *Placemaking: The Art and Practice of Building Community*. New York: Wiley.

Schneekloth, L. and Shibley, R. 2000. Implacing Architecture into the Practice of Placemaking. *Journal of Architectural Education*, 53(3), 130-40.

Seligman, M. 2003. *Authentic Happiness Using the New Positive Psychology to Realize Your Potential for Lasting Fulfilment*. London: Nicholas Brealey.

Taylor, C. 2007. *A Secular Age*. Cambridge, MA: Harvard University Press.

Torbert, W. and Associates. 2004. *Action Inquiry: The Secret of Timely and Transforming Leadership*. San Francisco: Berrett-Koehler.

Torbert, W. and Reason, P. (eds) 2001. 'Toward a participatory worldview: In physics, biology, economics, ecology, medicine, organizations, spirituality and everyday living'. *ReVision*, 23 (3-4)

Vernon, M. 2008. *Wellbeing*. Stocksfield: Acumen.

Wight, I. 2002. *Place, Placemaking and Planning: Part 1 – Wilber Integral Theory.* Paper to the Association of Collegiate Schools of Planning (ACSP) conference: Baltimore, November 2002.

Wight, I. 2005. Placemaking as Applied Integral Ecology: Evolving an Ecologically-Wise Planning Ethic. *World Futures*, 61, 127-37.

Wight, I. 2006. *Sensing Place Through an Integral Lens: Pointers for a Postmodern Planning as Place-making.* Paper to the 'Senses of Place' Conference: Hobart, Tasmania, April 2006.

Wight, I. 2011. Planning by Design in Community: Making Great Places? *Plan Canada*, 51(2), 29-32 Summer 2011.

Wilber, K. 1994. *Sex, Ecology and Spirituality.* Boston, MA: Shambhala.

Wilber, K. 1996. *A Brief History of Everything.* Boston, MA: Shambhala.

Wilber, K. 1997. *The Eye of Spirit: An Integral Vision for a World Gone Slightly Mad.* Boston, MA: Shambhala.

Wilber, K. 1998. *The Marriage of Sense and Soul: Integrating Science and Religion.* New York: Broadway Books.

Wilber, K. 2000a. *Integral Psychology: Consciousness, Spirit, Psychology, Therapy.* Boston, MA: Shambhala.

Wilber, K. 2000b. *A Theory of Everything: An Integral Vision for Business, Politics, Science and Spirituality.* Boston, MA: Shambhala.

Wilber, K. 2006. *Integral Spirituality: A Startling New Role for Religion in the Modern and Postmodern World.* Boston, MA: Integral Books.

Wilber, K. 2007. *The Integral Vision: A Very Short Introduction to the Revolutionary Integral Approach to Life, God, the Universe, and Everything.* Boston, MA: Shambhala.

Wilber, K., Patten, T., Leonard, A. and Morelli, M. 2008. *Integral Life Practice: A 21st-Century Blueprint for Physical Health, Emotional Balance, Mental Clarity and Spiritual Awakening.* Boston, MA: Integral Books.

Winikoff, T. (ed.) 1995. *Places Not Spaces: Placemaking in Australia.* Australian Council for the Arts: Sydney NSW Australia.

Index